THIS IS IMPROBABLE TOO

Synchronized Cows, Speedy Brain Extractors
and More WTF Research

Marc Abrahams

502
Abrahams

A Oneworld Book

First published in North America, Great Britain & Australia
by Oneworld Publications, 2014

Copyright © Marc Abrahams 2014

The moral right of Marc Abrahams to be identified as the Author of
this work has been asserted by him in accordance with the Copyright,
Designs and Patents Act 1988

ISBN 978-1-78074-361-5
Ebook ISBN 978-1-78074-362-2

Typesetting and eBook design by Tetragon
Cover design copyright © Jim Tierney
Printed and bound in Denmark by Nørhaven A/S

Illustration credits on pp. 315–17

Oneworld Publications
10 Bloomsbury Street, London WC1B 3SR, England

Stay up to date with the latest books,
special offers, and exclusive content from
Oneworld with our monthly newsletter

Sign up on our website
www.oneworld-publications.com

CONTENTS

BETWEEN THE SECOND AND FOURTH DIGITS

If you plunge or plod through these pages, expect the unexpected. I went to a lot of trouble to find it for you, and then worked to describe it simply and clearly – more clearly, in many cases, than it may have presented itself.

I collect and write about improbable research. Here's what those words mean to me. Improbable: not what you expect. Research: the attempt, intentional or not, to find or understand something that no one has yet managed to find or understand.

I do improbable research about improbable research.

Some of what I find goes into my 'Improbable Research' column in the *Guardian* newspaper. Some of it goes into the magazine I edit, the *Annals of Improbable Research*.

Some of it ends up earning an Ig Nobel Prize. I founded the Ig Noble Prize ceremony in 1991, and every year we (a shadowy group called the Ig Nobel Board of Governors) award ten new Ig Nobel Prizes for achievements that first make people laugh, then make them think.

That's the quality I always look for: that whatever the story is, it – with no twisting or adornment – first makes people laugh, then makes them think.

This book, *This Is Improbable Too*, is the second book in the series that began with *This Is Improbable*. That 'too' is meant to imply two things.

First, that this book is second.

And second, that the stories I write about do not stand alone – the people who did these things also did other things, some of which are fully as unexpected. It's easy to assume that the good story you know about a person is *the* good story about that person. In my experience, poking through studies and books, and chatting and gossiping with thousands of improbable people, if there's one good story about a person, chances are high that other stories exist too, and that some of those stories are even better than the one you knew about.

The stories in this book are all, one way or another, about people, arrayed somewhat by body part.

You might notice that two of those people keep reappearing.

One of those individuals began, in middle age, to count things that annoyed him. I don't mean by that that he keeps a long list of the many things that annoy him. No. This fellow, when he's bored enough, takes note of some particular thing that has repeatedly annoyed him. He then carefully counts how many times that annoying thing occurs during a particular span of time. Then he publishes a report about it, in some scholarly journal.

The other individual began, also in mid-adulthood, to pointedly find a connection between the relative length of a person's fingers and important aspects of that person's life. He also publishes his reports in scholarly journals.

The first of those individuals leans toward attributing no significance to what he sees. Bean-counting, done his way, is almost a form of poetry. To him, it's a source of grim, soul-satisfying amusement. Tally ho. Here are representative passages from his body of research:

- 'A total of 45 1-hr. citings of convenience were taken, during the Summer of 1983, equally divided among Tuesday, Wednesday, and Thursday, 1000 to 1500 hours. Note was made of the north-south movement, which remained approximately consistent during the period at a rate of about 800 private and 1,850 commercial vehicles per hour. The timing of the traffic signal was constant: 45 sec. "go" (green), 4 sec. "caution" (yellow), and 41 sec. "stop" (red). Notice was taken of the number of vehicles passing the stop light, where passing the stop light was defined as entering into and continuing through the intersection after the signal had turned red.'

- 'The students were asked for a single answer to the following query, in my opinion brussels sprouts are (a) very repulsive, (b) somewhat repulsive, (c) something that I can either take or leave, (d) somewhat delicious, or (e) especially delicious. The findings for the group (collectively) and (stratified) by sex and nationality are shown in Table 1.'

Table 1:

Number of Responses and Percent for Totality, by Sex, and by Citizenship

Response	Total group (n=442)		Women (n=266)		Men (n=176)		US national (n=217)		Foreign (n=225)	
	%	n	%	n	%	n	%	n	%	n
Very repulsive	31	137	30	80	34	60	40	87	22	50
Somewhat repulsive	20	88	20	53	19	33	19	41	20	45
Indifferent	41	181	43	114	38	67	36	78	46	104
Somewhat delicious	6	27	6	16	8	14	4	9	10	23
Especially delicious	2	9	1	3	1	2	1	2	2	3

The other individual leans towards attributing significance where someone else might see only fingers. This is a form of leadership, done so that others might perceive his insights. His jargon phrase '2D:4D' means 'the relative lengths of the second finger and the fourth finger':

- 'We recruited 300 subjects (117 men and 183 women) with a minimum age of 30 years from the Merseyside area. Participants were from social groups of elderly retired people and mature university students. We measured the 2nd and 4th digit length twice... The English sample... showed that married women had higher 2D:4D ratios than unmarried women.'

- 'We measured the lengths of the 2nd (index) and 4th (ring) finger in a sample of young men and recorded short digital video clips of their dance movements. A panel of 104 female judges rated 12 clips of men with the lowest and highest finger-length ratios (2D:4D) for attractiveness, dominance, and masculinity.'

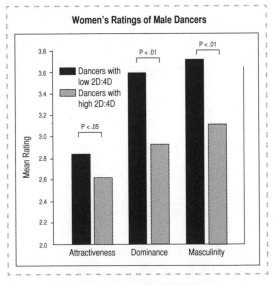

A classic in the body of 2D:4D work

I hope those quotes appeal to you enough, or perplex you enough, that you will track down the journals in which they appear, and find the bigger stories there. And I hope other parts in the body of this book have a similar effect. I've told you only a short version of each story. Still more juicy improbable details, unmentioned by me, await you. The references noted at the end of each story point you to treasures. (For the examples in this introduction, though, I leave you the pleasure of googling them to find the citations.)

Chunks of what's here appeared in the newspaper column. Chunks were in the magazine. Much of it came into existence with and for this book, updating or augmenting the newspaper or magazine chunks, or becoming wholly new bits of the universe.

The seven billion or so humans of planet Earth have been relentlessly kind in doing improbable things that deserve to be written up. I am way behind in that writing, and am relentlessly scrambling to try to catch up.

But if ever you find an especially good improbable thing that you wish someone would write about, I wish you would write to me about it. You may find me, amidst steepening heaps of improbable research, at www.improbable.com.

Sincerely and improbably,

Marc Abrahams

Editor and Co-founder,
Annals of Improbable Research

PS. What is the best way to read this book? I suggest that each night you choose a different story, and read it aloud to loved ones, at bedtime.

THIS IS IMPROBABLE TOO

ONE

THE BRAIN'S BEHIND

IN BRIEF

'WOULD BOHR BE BORN IF BOHM WERE BORN BEFORE BORN?'
by Hrvoje Nikolić (published in the *American Journal of Physics*, 2008)

Some of what's in this chapter: Dr Bean, man of international body parts • Einstein, Einstein, Einstein and other Einsteins • Improved scarecrow • The man who has Gorbachev's number • His basic laws of stupidity, and theirs of incompetence • Speaking of hooked tongue • Criminal mentors • When Washington really counted • The man who really counts: Trinkaus • Strange seats for prominent minds • Kakutani's bottled-up thoughts • Portfolio of a genius • One theory of everything • A number of genius numbering schemes

BEAN: COUNTER OF BODY PARTS

Dr Robert Bennett Bean took the measure of his fellow men almost fanatically. Women, too. He measured the parts, then published the copious details, and sometimes pictures, for all to see.

Bean worked at the University of Michigan, then at the Philippine Medical School, then at Tulane University, and finally at the University of Virginia. One of his first published papers, in 1907, was 'A Preliminary Report on the Measurements of about 1,000 Students at Ann Arbor, Michigan'. After that, he turned more specific, looking at this or that particular organ, limb or bodily region.

Bean measured lots of innards. In 'Some Racial Character-istics of the Spleen Weight in Man', he wrote: 'The white male spleen weighs about 140 grams, the negro male 115 grams, the white female 130 grams and the negro female 80 grams.' Num-bers abound also in his 'Some Racial Characteristics of the Liver Weight in Man', and 'Some Racial Characteristics of the Weight of the Heart and Kidneys'.

He occasionally looked at the entire person, as in 'Notes on the Hairy Men of the Philippine Islands and Elsewhere'.

Most often, though, he did piece work. In 'Sitting Height and Leg Length in Old Virginians', he instructed: 'The sitting height, leg length, and sitting height index of several groups of Old Virginians is of some interest.'

Bean's treatise on ears is divided into two parts: 'Ears of the morgue subjects' and 'Ears of the living subjects'.

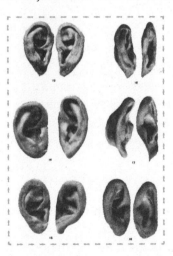

'Characteristics of the External Ear' collected by Robert Bennett Bean, including ears of a Filipino woman, a Filipino man and a Russian (gender unspecified)

He published 'Note on the Head Form of 435 American Soldiers with Special Reference to Flattening in the Occipital Region', and also 'Three Forms of the Human Nose'. Sometimes he was very specific: 'The Nose of the Jew and the Quadratus Labii Superioris' (facial muscle).

In 'Some Useful Morphologic Factors in Racial Anatomy', Bean introduced the omphalic index, a new metric about the belly button. One obtains it by making two measurements and a calculation: 'The distance of the umbilicus from the symphysis pubis is divided by the distance of the umbilicus from the suprasternal notch.'

By the time Bean died in 1944, he had recorded measurements of more partial people than almost anyone else ever had.

This was, obviously, not the same Dr Bennett Bean who, in 1980, published the study (described previously in *This Is Improbable*) entitled 'Nail Growth: Thirty-Five Years of Observation'. That was Robert Bennett Bean's son, William Bennett Bean, whose measurements were circumscribed, focusing exclusively on what he found at the ends of his own fingers.

Bean, Robert Bennett (1907). 'A Preliminary Report on the Measurements of About 1,000 Students at Ann Arbor, Michigan'. *Anatomical Record* 1: 67–8.

—, and Wilmer Baker (1919). 'Some Racial Characteristics of the Spleen Weight in Man'. *American Journal of Physical Anthropology* 2 (1): 1–9.

— (1919). 'Some Racial Characteristics of the Weight of the Heart and Kidneys'. *American Journal of Physical Anthropology* 2 (3): 265–74.

Bean, Robert Bennett (1913). 'Notes on the Hairy Men of the Philippine Islands and Elsewhere'. *American Anthropologist* 15 (3): 415–24.

— (1933). 'Sitting Height and Leg Length in Old Virginians'. *American Journal of Physical Anthropology* 17 (4): 445–79.

— (1915). 'Some Characteristics of the External Ear of American Whites, American Indians, American Negroes, Alaskan Esquimos, and Filipinos'. *American Journal of Anatomy* 18 (2): 201–25.

—, and Carl C. Speidel (1923). 'Note on the Head Form of 435 American Soldiers with Special Reference to Flattening in the Occipital Region'. *Anatomical Record* 25 (6): 301–11.

Bean, Robert Bennett (1913). 'Three Forms of the Human Nose'. *Anatomical Record* 7 (2): 43–6.

— (1913). 'The Nose of the Jew and the Quadratus Labii Superioris Muscle'. *Anatomical Record* 7 (2): 47–9.

— (1912). 'Some Useful Morphologic Factors in Racial Anatomy'. *Anatomical Record* 6 (4): 173–9.

Bean, William B. (1974). 'Nail Growth: Thirty-Five Years of Observation'. *Archives of Internal Medicine* 134 (3): 497–502.

Terry, R.J. (1946). 'Robert Bennett Bean, 1874–1944'. *American Anthropologist* 48 (1): 70–4.

IN BRIEF

'THE SPLENIC SNOOD: AN IMPROVED APPROACH FOR THE MANAGEMENT OF THE WANDERING SPLEEN'
by Steven P. Schmidt, H. Gibbs Andrews and John J. White
(published in *Journal of Pediatric Surgery*, 1992)

OTHER EINSTEINS

People say 'There is only one Einstein', but of course that is not so. Albert stands celebrated, but not alone.

Albert Einstein has a signature equation, $e=mc^2$, which predicts how energy relates to mass. M.E. Einstein of Purdue University in West Lafayette, Indiana, has a whole set of equations that predict the composition of a pork carcass.

M.E. Einstein and several collaborators published a series of studies – seven of them so far – in the *Journal of Animal Science*. Their 'Evaluation of Alternative Measures of Pork Carcass Composition' appeared in 2001. It is a minor classic in the history of pork-production prediction literature. This passage lists several of the parameters that Professor Einstein found ways to manipulate: 'FFLM is fat-free lean mass (kg), TOFAT is total carcass fat mass (kg), LFSTIS is lipid-free soft-tissue mass (kg), STLIP is soft-tissue lipid mass (kg), DL is dissected lean in the four lean cuts (kg), and NLFAT is the non-lipid components of the fat tissue.'

Evaluation of alternative measures of pork carcass composition[1,2]

A. P. Schinckel[3], J. R. Wagner[4], J. C. Forrest, and M. E. Einstein

M.E. Einstein also co-authored the doubly seminal 'Utilisation of a Sperm Quality Analyser to Evaluate Sperm Quantity and Quality of Turkey Breeders'. It was published in 2002 in the journal *British Poultry Science*.

Outside a small circle of specialists, Einstein's pork carcass composition equations and Einstein's turkey sperm quality analyser analysis are not so well known as they perhaps deserve to be.

Anyone with access to certain libraries can also check out Einstein on cannabis. Albert Einstein never published any research papers about cannabis, at least not formally, but Rosemarie Einstein did. In 1975, she and two colleagues at the University of Leeds investigated the use of cannabis – and alcohol and tobacco, too – by three hundred young persons at a university.

Einstein and her team carefully protected the students' confidentiality. In their study, which appeared in the *British Journal of Addiction*, no student is named. Even the university is not identified. The report speaks of it only as 'a provincial university', leaving readers to speculate, perhaps feverishly.

The scientists discovered exactly how many of those students used pot, alcohol, tobacco or any combination of the three. Or, to be more specific, they discovered what the students said they used. And how. According to the survey results, some students smoked their cannabis, others ate it, still others drank it. Some said they avoided cannabis altogether. Only a minority claimed to smoke tobacco, but none reported eating or drinking it. Almost everyone claimed to drink alcohol.

The scientists also discovered something they had expected: that students cannot be relied upon to answer surveys. The team says it sent questionnaires to exactly one thousand students, and that exactly three hundred of those questionnaires were returned. This 300/1,000 is a return rate of 33 percent, Einstein and her colleagues explain, using a brand of mathematics peculiarly their own.

There are many other Einsteins besides Albert, M.E. and Rosemarie. One analysed magical thinking in obsessive-compulsive persons. One did a comparison study of different kinds of barium enemas. One was a specialist in the history of television programmes. And so on. There is, I expect, an Einstein for everyone.

Schinckel, A.P., J.R. Wagner, J.C. Forrest and M.E. Einstein (2001). 'Evaluation of Alternative Measures of Pork Carcass Composition'. *Journal of Animal Science* 79 (5): 1093–119.

Schinckel, A.P., C.T. Herr, B.T. Richert, J.C. Forrest and M.E. Einstein (2003). 'Ractopamine Treatment Biases in the Prediction of Pork Carcass Composition'. *Journal of Animal Science* 81 (1): 16 Schinckel, A.P., 28.

Neuman, S.L., C.D. McDaniel, L. Frank, J. Radu, M.E. Einstein and P.Y. Hester (2002). 'Utilisation of a Sperm Quality Analyser to Evaluate Sperm Quantity and Quality of Turkey Breeders'. *British Poultry Science* 43 (3): 457–64.

Einstein, Rosemarie, Ian E. Hughes and Ian Hindmarch (1975). 'Patterns of Use of Alcohol, Cannabis and Tobacco in a Student Population'. *British Journal of Addiction to Alcohol & Other Drugs* 70 (2): 145–50.

Einstein, Danielle A., and Ross G. Menzies (2004). 'Role of Magical Thinking in Obsessive-Compulsive Symptoms in an Undergraduate Sample'. *Depression and Anxiety* 19 (30): 174–9.

Davidson, Jon C., David M. Einstein, Brian R. Herts, D.M. Balfe, Robert E. Koehler, Desiree E. Morgan, M. Lieber and Mark E. Baker (1999). 'Comparison of Two Barium Suspensions for Dedicated Small-Bowel Series'. *American Journal of Roentgenology* 172 (2): 379–82.

Einstein, Daniel (1997). *Special Edition: A Guide to Network Television Documentary Series and Special News Reports, 1980-1989*. Lanham, MD: Scarecrow Press.

AN IMPROBABLE INNOVATION

'A NEW AND USEFUL IMPROVEMENT IN SCARECROWS'
by Hugh Huffman and Ernest J. Peck (US patent no. 1,167,502, granted 1916)

The patent holders claimed that 'Prior to our invention, the scare crows ordinarily used were crude affairs... One of the main objects of our invention is to provide a more efficient form of scare crow'.

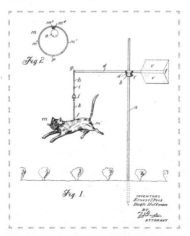

Illustrative diagram from 'A New and Useful Improvement in Scarecrows' from US Patent no. 1,167,502

PROBABILITY, BY GOD

In 1988, Robert W. Faid of Greenville, South Carolina, solved one of the oldest and most famous problems in mathematics. Yet almost no one noticed. Cracking the nut that was nearly two millennia old, Faid calculated the identity of the Antichrist.

In the rarified world of mathematicians, certain problems become the focus of intense pursuit. The Four-Colour Map Problem was finally solved, by Wolfgang Haken and Kenneth Appel, in 1976. Fermat's Last Theorem tantalized mathematicians until Andrew Wiles solved it in 1993.

Haken and Appel became instantly famous among mathematicians. Wiles became a worldwide celebrity.

But little academic or public acclaim came to Robert W. Faid, perhaps because no one had previously realized that the identity of the Antichrist *was* a mathematical problem.

The Antichrist problem has been on the books since about the year 90, when 'The Revelation of St John' brought it to public notice. Over the years, many amateur mathematicians joined the professionals in trying their hand at this delightful, yet maddening puzzle. Eventually it became a favourite old chestnut, something to be wondered at, but perhaps too difficult ever to yield a solution.

Then, after most had given up hope, Robert Faid solved it. In retrospect, his accomplishment seems almost absurdly simple: The Antichrist is Mikhail Gorbachev, with odds of 710,609,175,188,282,000 to 1.

There is no mystery to this. Faid is a trained engineer. He is methodical and rigorous. He wrote a book explaining every first and last tittle and jot: *Gorbachev! Has the Real Antichrist Come?*, published by Victory House. It tells where each number comes from and how it enters into the calculation. Professional mathematicians find it difficult to argue with the logic.

Outside the maths community, the book received little

attention, but Robert. W. Faid was nonetheless awarded with the Ig Nobel Prize in mathematics in 1993 for his achievement.

More recently, another good and great mathematical problem was knocked off. Stephen D. Unwin wrote a book called *The Probability of God*. It is much celebrated.

Stephen D. Unwin has a PhD in theoretical physics. Like Robert Faid, he has methodically, rigorously and with faithful certainty chosen some numbers, then performed addition, subtraction, multiplication and division. The calculated result: that there is almost exactly a 67-percent probability that God exists. The book reveals all the technicalities and includes a handy spreadsheet for those anxious to try the calculations for themselves. After following his detailed instructions for using Microsoft Excel to replicate the maths, he notes: 'You are now a mathematical theologist and can do things of which Aristotle, St. Thomas, and Kant only dreamed. Please proceed responsibly.' Like all good statistical reports, he does point out the possibility that something is off. There is, Stephen D. Unwin carefully warns us, a 5-percent chance that his calculation is wrong.

Faid, Unwin and God knows how many others give mathematicians faith that every problem, no matter how hard, can have some kind of devilishly simple solution.

Faid, Robert W. (1988). *Gorbachev! Has the Real Antichrist Come?* Tulsa, OK: Victory House. Unwin, Stephen D. (2003). *The Probability of God: A Simple Calculation That Proves the Ultimate Truth*. New York: Crown Forum.

MAY WE RECOMMEND

'THE DESK OR THE BED?'

by Robert Gifford and Robert Sommer (published in *Personnel and Guidance Journal*, 1968)

The authors, at the University of California at Davis and supported in part by a grant from the US Office of Education, concluded: 'There is nothing in these data to support the recommendations for studying in a straight-backed chair at a desk.'

THE BASIC LAWS OF HUMAN STUPIDITY, OR: THE GIFT OF INCOMPETENCE

The basic laws of human stupidity are ancient. The definitive essay on the subject is younger. Called *The Basic Laws of Human Stupidity*, it was published in 1976 by an Italian economist.

Professor Carlo M. Cipolla taught at several universities in Italy and for many years at the University of California, Berkeley. He also wrote books and studies about clocks, guns, monetary policy, depressions, faith, reason and of course – he being an economist – money. His essay about stupidity encompasses all those other topics, and perhaps all of human experience.

Cipolla wrote out the laws in plain language. They are akin to laws of nature – a seemingly basic characteristic of the universe. Here they are:

1) Always and inevitably, everyone underestimates the number of stupid individuals in circulation.

2) The probability that a certain person be stupid is independent of any other characteristic of that person.

3) A stupid person is a person who causes losses to another person or to a group of persons, while himself deriving no gain and even possibly incurring losses.

4) Non-stupid people always underestimate the damaging power of stupid individuals. In particular, non-stupid people constantly forget that at all times and places and under any circumstances to deal and/or associate with stupid people always turns out to be a costly mistake.

Cipolla's essay gives an X-ray view of what distinguishes countries on the rise from those that are falling.

Countries moving uphill have an inevitable percentage of

stupid people, yes. But they enjoy 'an unusually high fraction of intelligent people' who collectively overcompensate for the dumbos.

Declining nations have, instead, an 'alarming proliferation' of non-stupid people whose behaviour 'inevitably strengthens the destructive power' of their persistently stupid fellow citizens. There are two distinct, unhelpful groups: 'bandits' who take positions of power that they use for their own gain; and people out of power who sigh through life as if they are helpless.

Cipolla died in 2000, just a year after two psychologists at Cornell University in New York State wrote a study entitled 'Unskilled and Unaware of It: How Difficulties in Recognizing One's Own Incompetence Lead to Inflated Self-Assessments'. Without mentioning any form of the word 'stupidity', it serves as an enlightening and dismaying supplement to Cipolla's basic laws.

In the Cornell study, David Dunning and Justin Kruger supplied scientific evidence that incompetence is bliss, for the incompetent person. They staged a series of experiments, involving several groups of people. Beforehand, they made some predictions, most notably that:

1)   Incompetent people dramatically overestimate their ability; and
2)   Incompetent people are not good at recognizing incompetence – their own or anyone else's.

In one experiment, Dunning and Kruger asked sixty-five test subjects to rate the funniness of certain jokes. They then compared each test subject's ratings of the jokes with ratings done by eight professional comedians. Some people had a very poor sense of what others find funny – but most of those same individuals believed themselves to be very good at it, rather like David Brent of the television comedy *The Office*.

Another experiment involved logic questions from law

school entrance exams. The logic questions produced much the same results as the jokes. Those with poor reasoning skills tended to believe they were as good as Sherlock Holmes.

Overall, the results showed that incompetence is even worse than it appears to be, and forms a sort of unholy trinity of cluelessness. The incompetent don't perform up to speed; don't recognize their lack of competence; and don't even recognize the competence of other people.

Dunning explained why he took up this kind of research: 'I am interested in why people tend to have overly favorable and objectively indefensible views of their own abilities, talents, and moral character. For example, a full 94% of college professors state that they do "above average" work, although it is statistically impossible for virtually everybody to be above average.' In 2008, he and his colleagues revisited their findings with 'Why the Unskilled Are Unaware: Further Explorations of (Absent) Self-insight among the Incompetent', in order to show that their assessment was not a statistical artifact.

Participants—A total of 46 participants were recruited at a Trap and Skeet competition in exchange for a payment of $5. Most participants reported owning at least one firearm (96%) and having taken a course in firearm safety (89%). They possessed between 6 and 65 years experience with firearms (mean = 34.5 years).

Participants—Participants were 42 undergraduates who participated in exchange for extra credit in undergraduate psychology courses.

Participants in Study 3 (top) and Study 5 (bottom) to understand 'Why the Unskilled Are Unaware'

If you have colleagues who are incompetent and unaware of it, Dunning and Kruger's research is a useful and convenient tool. I recommend that you make copies of their reports, and send them – anonymously, if need be – to each of those individuals. (Professor Cipolla used that same method, minus the anonymity, to distribute his essay *The Basic Laws of Human Stupidity* among his closest friends.)

A copy might, too, be a helpful gift for any national or other leader to whom it may pertain.

For celebrating incompetence and unawareness, Dunning and Kruger won the 2000 Ig Nobel Prize in the field of psychology.

Cipolla, Carlo M. (1976). *The Basic Laws of Human Stupidity*. Bologna: The Mad Millers/ Il Mulino.

Dunning, David, and Justin Kruger (1999). 'Unskilled and Unaware of It: How Difficulties in Recognizing One's Own Incompetence Lead to Inflated Self-Assessments'. *Journal of Personality and Social Psychology* 77 (6): 1121–34.

Ehrlinger, Joyce, Kerri Johnson, Matthew Banner, David Dunning and Justin Kruger (2008). 'Why the Unskilled Are Unaware: Further Explorations of (Absent) Self-insight among the Incompetent'. *Organizational Behavior and Human Decision Process* 105 (1): 98–121.

IN BRIEF

> **A Story About a Stupid Person Can Make You Act Stupid (or Smart): Behavioral Assimilation (and Contrast) as Narrative Impact**
>
> MARKUS APPEL
> *Department of Education and Psychology, Johannes Kepler University of Linz, Linz, Austria*

'A STORY ABOUT A STUPID PERSON CAN MAKE YOU ACT STUPID (OR SMART): BEHAVIORAL ASSIMILATION (AND CONTRAST) AS NARRATIVE IMPACT'
by Markus Appel (published in *Media Psychology*, 2011)

MAY WE RECOMMEND

'A LUCKY CATCH: FISHHOOK INJURY OF THE TONGUE'
by Karen A. Eley and Daljit K. Dhariwal (published in *Journal of Emergencies, Trauma, and Shock*, 2010)

DASTARDLY DEVELOPMENT

Is our criminals learning?

The question is a natural follow-on to one raised by George W. Bush during his first campaign to become president of the

United States. On 11 January 2000, looking down at a select audience in the city of Florence, South Carolina, where the crime rate was 3.4 times the national average, Bush asked: 'Is our children learning?'

For Bush, it seemed, learning was a lifelong challenge. In the journal *Criminology,* Carlo Morselli and Pierre Tremblay, of the Université de Montréal, and Bill McCarthy, of the University of California at Davis, explore how that challenge applies to 268 prison inmates in the Canadian province of Quebec. Their report, called 'Mentors and Criminal Achievement', echoes the thoughts and findings not only of George W. Bush, but also of earlier researchers and criminals.

They offer up a nugget from Indiana University criminologist Edwin H. Sutherland's 1937 book *The Professional Thief, By a Professional Thief.* 'Any man who hits the big-time in crime, somewhere or other along the road, became associated with a big-timer who picked him up and educated him', the thief told Sutherland, adding: 'No one ever crashed the big rackets without education in this line.'

Mentors, say those who study the development of great executives, inventors, artists, sports figures and entrepreneurs, are crucial if one is to have a successful career. But aside from those highly celebrated professions, and from some obvious high-skill specialties, do people really need mentors or can they generally find success on their own? Do mentors make a measurable difference?

'Our analysis', write Morselli et al., 'focuses on the effects of mentors on two aspects of criminal achievement: illegal earnings and incarceration experiences … Protégés with lower self-control attract the attention of some criminal mentors who provide the structure and restraint that lead to a more prudent approach to crime. This approach involves fewer and more profitable offenses that lower the risks of apprehension and, perhaps, promote long-term horizons in crime.'

The researchers used a painstaking protocol: 'We collected information on monthly illegal earnings and on the number of days that respondents were incarcerated. After calculating the total for criminal earnings and incapacitation experiences for the period, we applied logarithmic transformations to create our dependant variables.'

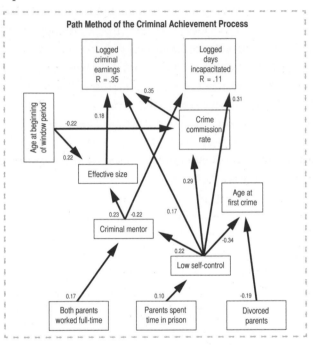

The authors note that 'For clarity... age at first crime on criminal earnings... and parents' full-time employment... were removed from the model.'

Their calculation resulted in a big payoff. As they put it: 'Our findings suggest that strong foundations in crime offer an advantageous position for continuous achievement and the presence of a criminal mentor is pivotal for achievement over one's criminal career.'

Morselli, Carlo, Pierre Tremblay, and Bill McCarthy (2006). 'Mentors and Criminal Achievement'. *Criminology* 44 (1): 17–43.
Sutherland, Edwin H. (1937). *The Professional Thief, By a Professional Thief.* Chicago: University of Chicago Press.

THE PRESIDENT WHO COUNTED

'Of all the presidents we have had, George Washington was the only who really counted.' That single sentence with its double meaning comes at the end of a four-page monograph published in 1978 in the *Alabama Journal of Mathematics*. The report's title is 'George Washington: He Liked to Count Things'. The author, Pete Casazza, a mathematician who writes under multiple pen names, adds fuel to a tiny fire of his own creation, suggesting that the number one number one official of the United States of America was a little obsessed. (Casazza, now a maths professor at the University of Missouri, wrote this under the pseudonym Cora Green, back when he was at Auburn University in Montgomery, Alabama.)

Casazza/Green specifies more than forty specific things that America's first president counted. They are mere examples, he emphasizes, plucked from a myriad: 'First, and foremost, he liked to count things on his plantation at Mount Vernon. He counted and recorded his horses, cataloging them by color, working mares and others, unbroken or not, as well as recording their height, age and weight. He counted ewes, hogs, calves, yearlings, spades, axes, and knives.'

In war time, as commander of the rebel American army, General Washington counted 'soldiers and armies (as well as the distances between them), guns, ships, horses, mortars, batteries … the number of casualties suffered by his army … listing time periods, killed and wounded, and separating it into colonels, Lt colonels, majors, captains, lieutenants, sergeants, and privates'. In peace time, President Washington counted how many bushels of wheat were sown on his farm and how many trees – oak, yew, hemlock, aspen, magnolia, elm, papaw, lilacs, fringe, swamp berry – were planted.

He counted nuts. He counted seeds. He counted miles travelled, and compared them with the sometimes inaccurate distances marked on maps.

President Washington's diary of Sunday, 11 May 1788, when he
'counted the number of the following articles'. He says he was
'home all day'.

Washington was originally a surveyor by profession, and
perhaps also by inclination. Casazza/Green writes: 'He recorded
in his diary on May 11, 1788, that he had spent the entire Sunday
at home counting different kinds of peas and beans … He found
it took exactly 3,144 of the small round peas known as gentle-
man's peas to fill a pint, 2,268 peas of the kind he brought from
New York, 1,375 of the peas he had brought from Mrs. Danger-
field's, 1,330 of those he had been given by Heziah Fairfax, 1,186
of the large black-eyed peas, and 1,473 bunch hominy beans.

Having arrived at his count, he next calculated the number of hills a bushel of each kind of peas and the beans would plant, allowing five to a hill.'

The monograph supplies no clear evidence as to whether Washington actually enjoyed totting. Who, under any circumstances, can say for sure what was in another person's mind? But neither does it try to persuade us that this was a psycho-medical problem, an undiagnosed case of obsessive-compulsive disorder. And for that, we might choose to count our blessings.

Green, Cora (1978). 'George Washington: He Liked to Count Things'. *Alabama Journal of Mathematics* 2 (2): 43–6.

RESEARCH SPOTLIGHT

'PRESIDENT OBAMA'S CORONARY CALCIUM SCAN'
by Andrew J. Einstein (published in the *Archives of Internal Medicine*, 2010)

TRINKAUS: AN INFORMAL LOOK

'Too little known' is the common verdict about John W. Trinkaus's research. Some say this of his discoveries, which are proudly and surprisingly unsurprising. Most, though, are referring to the very existence of Trinkaus's published reports, which are numerous (nearly one hundred so far, with more on the way) and pithy (with few exceptions, each is a page or two in length), and which concern a wide range of common behaviour that Dr Trinkaus finds annoying or anomalous.

For decades, Dr Trinkaus, at and then retired from the Zicklin School of Business in New York City, has conducted research on attitudes about Brussels sprouts; on the marital status of television quiz show contestants; on bakery wrapping-tissue usage; and on how many people wear baseball-type caps with the bill facing backwards. And on many other things. What percentage of pedestrians wear sports shoes that are white rather than some other colour. What percentage of swimmers swim laps in the shallow end of a pool rather than the deep end. What percentage of car drivers almost, but not completely, come to a stop at one particular stop sign. What percentage of commuters carry attaché cases. What percentage of shoppers exceed the number of items permitted in a supermarket's express checkout lane.

To sense the flavour of Trinkaus's research and writing, one might dip into his somewhat celebrated Brussels sprouts paper. Published in 1991 in the journal *Psychological Reports*, it is entitled 'Taste Preference for Brussels Sprouts: An Informal Look'. The professor writes: 'As to the apparent greater acceptability of Brussels sprouts by older students, two possible explanations may be suggested. First, being older, they may have more experience with the vegetable, for example, having actually tried it, rather than classifying it as repugnant simply because of its name or reputation. Second, being older, they may have eaten more Brussels sprouts and found that after a while they began to like the taste.' (This, by the way, is one of Trinkaus's few co-authored papers. Generally, the man went solo.)

For a full appreciation of Trinkaus's body of work, one must read the original reports in their full detail. For those who have yet to enjoy that experience, we provide a somewhat haphazard index of a haphazard sampling of his work:

- Baseball-type caps, manner of wearing
- Brussels sprouts, taste preference for
- Business students' feelings about the academy and themselves
- Mobile phones, drivers' use of
- Multiple choice test questions, opinions about whether to change answers of
- Queasiness
- Railroad departure service, varying quality of

- Buzzwords in political commercials
- Doors, open versus closed, behaviour of peole regarding
- Elevators, communications in
- Expensive versus inexpensive cars, behaviour of owners of
- Fads of yesteryear
- Fruits and vegetables, business students' consumption of
- Gloves, lost
- Husbands with working wives
- Integrity, the word
- Looks, another
- Look, final
- Look, follow-up
- Look, further
- Look, an informal
- Manners, business
- Santa, children's perceptions about visiting
- Santa, parents' perceptions about visiting
- Sirens, use of
- Sports shoes, colour preference for
- Supermarket checkout, counting of items at
- Supermarket checkout, delays at
- Uncooked minced beef, shoppers' perceptions about
- Vision tests for driver licensing, honesty of people taking
- Winter storms, weather persons' predictions of
- Work arrival time
- Work departure time
- Women driving vans
- Yes, the word

While John W. Trinkaus sometimes suggests possible ways to look at his findings, interpretation is something he mostly – and proudly – avoids. When he notices something that can be tallied, he tallies it. In a profession ruled by the famous dictum 'publish or perish', he counts.

Trinkaus has become at least slightly better known since he was awarded the Ig Nobel Prize in literature in 2003; he, together with his research, was symbolically donated (by this author) to the BBC radio programme *The Museum of Curiosity* in 2012. On 10 May 2013, Professor Trinkaus gave his last lecture at the Zicklin School, after fifty years on the faculty.

Trinkaus, John, and Karen Dennis (1991). 'Taste Preference for Brussels Sprouts: An Informal Look'. *Psychological Reports* 69 (3): 1165–6.
Kaswell, Alice Shirrell (2003). 'Trinkaus: An Informal Look'. *Annals of Improbable Research* 9 (3): 4–15, http://www.improbable.com/airchives/paperair/volume9/v9i3/trinkaus0.html.

CALL FOR INVESTIGATORS

The Strange Seat Taxonomy, announced here, is searching for specimens of unusual academic chairs (in other, grander, words: endowed faculty positions, which in many cases do include actual pieces of furniture on which the faculty member can physically sit).

Definition: For the purposes of the project, a Strange Seat is an actual academic chair at a university or college that is memorable or unusual or that has a particularly curious cogno-intellectual story behind it. As a model specimen, consider: The Streisand Professorship in Intimacy and Sexuality at the University of Southern California, generously funded by the Funny Girl-cum-Funny Lady herself since 1984.

Purpose: To identify and catalogue the chairs and, were applicable, their endowments.

If you know of a unique specimen to add to the collection, please provide:

1. *The name and a twenty-word description of the chair, unless the name of the chair is twenty words longer, in which case the description should be reduced by a factor.*
2. *A photograph of the current chair, if available.*
3. *Several Internet URLs pointing to clear, unarguable documentation of the chair's existence at the assigned institution of higher learning.*

Send to marca@improbable.com with the subject line:
STRANGE SEAT TAXONOMY

FLOATING IDEAS

Almost nothing is more romantic than a mathematical theorem – if that theorem is stuffed into a bottle and cast adrift during a perilous sea voyage in wartime, and if the person who wrote it is one of the world's top mathematicians. Shizuo Kakutani, who died in 2004, just a few days before his ninety-third birthday, threw many such bottles – call them hundred-proofs, if you will – into the ocean when he was a young man. The fate of those bottles is a complete mystery.

Kakutani went on to become a legendarily great mathematician. Like most famous mathematicians, his fame is mostly limited to those in his profession.

Indirectly, though, the public is almost aware of Kakutani for two reasons. The book *A Beautiful Mind* was about the mathematician John Nash, who won a Nobel Prize in economics. Nash's most famous concept, a Nash Equilibrium, is based on the Kakutani Fixed Point Theorem. And Shizuo Kakutani's daughter, Michiko, is a Pulitzer Prize-winning book reviewer for the *New York Times*.

Although Shizuo Kakutani was a gregarious sort for a mathematician, the story of his bottled theorems was only told outside the tight circles of those who really, truly, deeply understand the nature of, well, circles, shortly after his death. Stanley Eigen, a mathematics professor at Northeastern University in Boston, wrote an appreciation of his longtime collaborator and friend for the *Annals of Improbable Research*. Eigen explains:

> At the start of World War II, Kakutani was a visiting professor at the Institute for Advanced Study in Princeton. With the outbreak of war he was given the option of staying at the Institute or returning to Japan. He chose to return because he was concerned about his mother.
>
> So he was put on a Swedish ship which sailed across the Atlantic, down around the Cape, and up to Madagascar, or thereabouts, where he and other Japanese were traded for Americans aboard a ship from Japan.
>
> The trip across the Atlantic was long and hard. There was the constant fear of being torpedoed by the

Germans. What, you may wonder, did Kakutani do. He proved theorems. Every day, he sat on deck and worked on his mathematics. Every night, he took his latest theorem, put it in a bottle and threw it overboard. Each one contained the instruction that if found it should be sent to the Institute in Princeton. To this day, not a single letter has been received.

Are some of those missing proofs startling and important? Is there any real chance of finding them? No one knows.

There is precious little scholarship about messages found in bottles. Devoted scavengers will find three little piles. Robert Kraske's too-slim (at only ninety-six pages) book *The Twelve Million Dollar Note: Strange But True Tales of Messages Found in Seagoing Bottles*. The messages-in-bottles collection at the Turks and Caicos National Museum. The rubber-duckies-and-other-things-that-wash-up-on-beaches research of Seattle-based oceanographer Curtis Ebbesmeyer. These, our greatest chronicler/gatherers of such materials, have so far disappointed us in the case of Kakutani.

Still, still, in the combined vastness of the oceans and of time, some of Kakutani's theorems might come into the possession of persons who can both read and appreciate their value.

Kraske, Robert (1977). *The Twelve Million Dollar Note: Strange But True Tales of Messages Found in Seagoing Bottles* Nashville: T. Nelson.
Sadler, Nigel (2001). 'Essay on Messages Found in Bottles'. *The Astrolabe: The Newsletter of the Turks and Caicos National Museum* (autumn): n.p.

PORTFOLIO OF A GENIUS

I'd been awaiting the arrival of the latest edition of *Portfolio of a Genius*. For the better part of fifteen years, I had been receiving the laboriously crafted, increasingly thick versions of this wondrous work. They arrived in my mailbox, always surprising by their very existence.

The author, James E. Shepherd Jr – the subject and author of the *Portfolio* – switched from paper to CD just before the turn of the millennium, perhaps at the request of the heavily burdened postal workers of the world. Each new paper version was thicker than its predecessor, and weightier, too. 'Mighty thick and mighty heavy' would be a good way to describe the later pre-CD incarnations.

The CD versions were of course svelter, but also fuller than ever with documentation of the life, the correspondence and especially the correspondence about the correspondence, of Mr Shepherd. Each new version contained all that was in its predecessors, and also copies of all subsequent correspondence sent and received pertaining thereto. In the 2009 edition, Shepherd included a photo of his MENSA membership card, which was due to expire on 31 March of that year. It was a new item in the *Portfolio*.

A web version existed for several years, but then vanished. I am intending to schedule time to schedule time to begin reading a new version, in whatever form it may take, if and when it appears. Perhaps you will, too.

Shepherd is a nonpareil – but he is not the only genius who has no match or equal, and no real rival. Somehow, though each is unique, they are legion. They seem, most of them, not to acknowledge, or maybe not even to see, the presence of the others. It's almost as if each exists in his or her own universe. Nearly every one of these peerless peers has his or her own theory about their particular universe, and about how and why he or she is peerless. In my experience, with what seems a large number of such geniuses, hes outnumber shes.

Here is a sampling of their books. The public may scoff, but it's possible that somewhere in the midst of this list is a theory that really does explain everything.

Theory of Interaction: The Simplest Explanation of Everything by Eugene Savov.

A Theory of Everything: An Integral Vision for Business, Politics, Science, and Spirituality by Ken Wilber. Wilber writes that 'the leading edge of consciousness evolution stands today on the brink of an integral millennium ...'

And Now the Long Awaited – "THEORY OF EVERYTHING" by Eugene Sittampalam. The book's back cover makes a calculatedly interesting offer: 'The author welcomes a refutation from any reader and offers hereby all his profits from this work – UP TO ONE MILLION US DOLLARS – to the first reader to successfully do so.'

The Theory of Everything: The Origin and Fate of the Universe by Stephen W. Hawking

The Power and Freedom of the Human Spirit: Introducing Another Theory of Everything by Earle Josiah, offers 'A coherent theory of everything that presents a new understanding of the wondrous world that extends beyond the barriers of normal life.'

Unitivity Theory by Leroy Amunrud claims that: 'To the knowledge of the author, this is the first book to contain a derived theory that treats the whole universe in a unified way, and discloses that the universe is simple as one, two, four.' The author, who says he previously worked on the Apollo Project at Honeywell Inc and Montana State University, has also 'been involved in ranching through out his entire life where he studied the logic of the cowboys'.

The Scientific Theory of Everything by Pacifico M. Icasiano has a cover photo of Mr Icasiano holding his chin. The publisher says, 'Pacifico M. Icasiano presents *A Scientific Theory of Everything* and unifies all sciences, both physical and metaphysical, including religion!'

There are many other such books. In theory – and, I expect, in fact – there will be many more.

Savov, Eugene (2002). *Theory of Interaction: The Simplest Explanation of Everything*. Sofia, Bulgaria: Geones Books.

Wilber, Ken (2000). *A Theory of Everything: An Integral Vision for Business, Politics, Science, and Spirituality*. Boston: Shambhala.

Sittampalam, Eugene (2008). *And Now the Long Awaited – "Theory of Everything"*. New York: Vantage Press.

Hawking, Stephen W. (2002). *The Theory of Everything: The Origin and Fate of the Universe*. Beverly Hills, CA: New Millennium Press.

Josiah, Earle (2007). *The Power and Freedom of the Human Spirit: Introducing Another Theory of Everything*. Charleston, SC: BookSurge Publishing.

Amunrud, Leroy (2007). *Unitivity Theory: A Theory of Everything*. Bloomington, IN: AuthorHouse.

Icasiano, Pacifico M. (2003). *The Scientific Theory of Everything*. Pittsburgh, PA: Dorrance Publishing.

MAY WE RECOMMEND

'AN EXCEPTIONALLY SIMPLE THEORY OF EVERYTHING'
by Antony Garrett Lisi (published in arXiv, 2007)

The author finishes his detailed report by stating that if his 'theory is fully successful as a theory of everything, our universe is an exceptionally beautiful shape'.

BRILLIANT EARLY EXPLANATIONS OF BRILLIANCE

Psychologists still grind away (sometimes at each other) at explaining what genius is, and where it comes from. The effort, now weary and tendentious, was exciting in its earlier days. In 1920, Lewis Terman and Jessie Chase of Stanford University published a report called 'The Psychology, Biology and Pedagogy of Genius', summarizing all the important new literature on the subject.

Those early twentieth-century psychologists showed a collective genius for disagreeing about almost everything.

J.C.M. Garnett, in a study called 'General Ability, Cleverness, and Purpose', offered a formula for genius. Measure a person's general ability; then measure their cleverness, then square both numbers and add them together, then take the square root. Genius.

We learn about C.L. Redfield, who 'cites 571 specially selected pedigrees to prove his theory' that 'rapid breeding inevitably leads to the production of inferior stock', but that 'inferior stock can be transformed into superior stock in 100 years, and into eminent men in 200 years.'

James G. Kiernan wrote a monograph called 'Is Genius a Sport, a Neurosis, or a Child Potentiality Developed?' Terman and Chase tell us that 'Kiernan, after a description of the ability of various men of genius, arrives at the conclusion that genius is not a sport nor a neurosis.' Kiernan's paper hints, right at the start, that its author knew neurosis intimately. The byline lists a few of his credentials, beginning with: Fellow, Chicago Academy of Medicine; Foreign Associate Member, French Medico-Psychological Association; Honorary Member, Chicago Neurologic Society; Honorary President, Section of Nervous and Mental Disease Pan-American Congress; Chairman, Section on Nervous and Mental Diseases, American Medical Association; and continuing on at some length.

A book by Albert Mordell explains that 'the literary genius is one who has experienced a repression, drawn certain conclusions from it, and expressed what society does', and that 'By making an outlet for their repressions in imaginative literature Rousseau, Goethe and many others have saved themselves from insanity.'

Bent on being thoroughly inclusive, Terman and Chase mention a book called *Jesus, the Christ, in the Light of Psychology*, by G. Stanley Hall. 'In two volumes', they write, 'Hall has given us an epochmaking study, chiefly from the psychological point of view, of the greatest moral genius of all time.' Terman and Chase seem to carefully dodge a bullet (or maybe a firing squad or even a massive artillery bombardment) of criticism, remarking only that 'It is impossible even to characterize such a monumental work in the few lines here available, much less to summarize it'. There's much more.

All told, Terman and Chase describe ninety-five scholarly and semi-scholarly papers and books, devoting a sentence or three to each of them. The exception, the lengthiest section of their report, is a lavish description of Terman's own recent studies, commencing with the words 'Terman devotes 102 pages of his latest book to . . . '. Terman's writings, reportedly, are filled with insights 'of special interest'.

Terman, Lewis M., and Jessie M. Chase (1920). 'The Psychology, Biology and Pedagogy of Genius'. *Psychological Bulletin*, 17 (12): 397–409.

Garnett, J.C.M. (1919). 'General Ability, Cleverness, and Purpose'. *British Journal of Psychology* 9: 345–66.

Redfield, Casper Lavater (1915). *Great Men and How They Are Produced*. Chicago: Privately pub., p. 32.

Kiernan, James G. (1915). 'Is Genius a Sport, a Neurosis, or a Child Potentiality Developed?'. *Alienist and Neurologist* 36: 165–82, 236–46, 384–95.

— (1916). 'Is Genius a Sport, a Neurosis, or a Child Potentiality Developed?'. *Alienist and Neurologist* 37: 70–82, 141–57.

— (1919). 'Is Genius a Sport, a Neurosis, or a Child Potentiality Developed?'. *Alienist and Neurologist* 40: 114–18.

Mordell, Albert (1919). *The Erotic Motive in Literature*. New York: Boni & Liveright, p. 250.

Hall, G. Stanley (1917). *Jesus, the Christ, in the Light of Psychology*. New York: Doubleday & Page, p. 733.

IN BRIEF

'IMAGES OF MADNESS IN THE FILMS OF WALT DISNEY'
by Allen Beveridge (published in *Psychiatric Bulletin*, 1966)

PSYCHIATRY AND THE MEDIA

Images of madness in the films of Walt Disney

Allan Beveridge

It has been demonstrated that images of madness in the media powerfully influence the general public and serve to perpetuate popular stereo-

gossipy elephants, jeering little boys, an ineffectual ringmaster and drunken clowns. To be excluded from this society would almost seem to

TWO

HIGH, BRAINY MINDEDNESS

IN BRIEF

**'"HOLY PTSD, BATMAN!": AN ANALYSIS OF THE PSYCHIAT-
RIC SYMPTOMS OF BRUCE WAYNE'**
by S. Taylor Williams (published in *Academic Psychiatry*,
2012)

*Some of what's in this chapter: How to grasp Machiavellianism •
Mental magic • One: Must take what they say seriously • Spatula to
the brain • Honestly: Appropriated thoughts • Judging while asleep •
Strudels in the brain • Laugh into unconsciousness • Mutter, Russia,
about unhappiness • Prozac is for animals • In defence of one's new
parking spot • Red bull bull • An Eiffel Tower in the head • Insulting
insulting language • Bad breath in the head*

PRINCELY BEHAVIOUR

By reputation, stockbrokers have manipulative personalities.
So do people who sell cars or buildings. Professor Abdul Aziz
took the measure of these groups of professionals, hoping to
see whether each lives up or down to the legend.

Aziz, who teaches business at Morgan State University in
Baltimore, Maryland, together with colleagues published three
studies a decade ago: 'Relations of Machiavellian Behavior with
Sales Performance of Stockbrokers'; 'Machiavellianism Scores
and Self-rated Performance of Automobile Salespersons'; and
'Relationship between Machiavellianism Scores and Perfor-

mance of Real Estate Salespersons'. All appear in the journal *Psychology Reports*.

Aziz explains that a Machiavellian person is someone who 'views and manipulates others' for 'personal gain, often against the other's self-interest'. He says this 'modern concept of Machiavellianism was derived from the ideas of [Niccolò] Machiavelli as published in [his book] *The Prince* in 1532', and that interest in it as a personality trait blossomed in the 1970s.

Aziz used a questionnaire based on psychological tests devised in the 1960s that claim to measure Machiavellianism by presenting statements and asking the test-taker to agree or disagree. The statements range from the goody-goody: 'Most people who get ahead in the world lead clean, moral lives', to the not-so-goody: 'The biggest difference between most criminals and other people is that the criminals are stupid enough to get caught'.

for personal gain, often against the other's self-interest'. Christie and Geis (1970) also developed scales called Mach IV and Mach V to measure Machiavellianism. These scales were based on three factors including tactics, views of human nature, and morality derived from Machiavelli's writings (Christie & Geis, 1970, p. 17). Those with high scores on a Machiavellianism scale have been called Machiavellian or high Machs; those with low scores on the scale have been called non-Machiavellian or low Machs (Christie & Geis,

From 'Relations of Machiavellian Behavior with Sales Performance of Stockbrokers'

Aziz prepared similar questions.

He got answers from 110 brokers who sell stocks on a commission basis. Aziz also wanted to know how good these stockbrokers were at their sales work, so he asked them to compare their own sales performance to that of their colleagues. Aziz would have preferred not to take the brokers' word for this. But, he writes, 'the company was not willing to disclose the actual amount of sales by individual stockbrokers'. After analysing what the stockbrokers told him, Aziz reports a strong association between the brokers' 'Machiavellian behaviour scale' rank and how good they claim to be at selling.

His conclusion: the stockbroker data support the 'assumption of a positive relationship between Machiavellianism and sales performance'.

Aziz then did a similar study of eighty car salespersons, all of whom work on commission. He asked them his Machiavellianism survey questions. He also asked each to tell him '(a) the number of cars sold during the previous year and (b) the income bracket that most closely matched their income during that year'. His conclusion: what the car salespersons told him provides 'partial support for earlier findings'.

Rounding out the Big Three, Aziz then talked with seventy-two estate agents who earn their money selling property on commission. The things they told him, Aziz says, 'support earlier results from samples of stockbrokers and automobile salespersons'.

A few other studies have cited Aziz's work. One of the first was a Canadian report called 'Psychopathy and the Detection of Faking on Self-Report Inventories of Personality'.

Aziz, Abdul, Kim May and John C. Crotts (2002). 'Relations of Machiavellian Behavior with Sales Performance of Stockbrokers'. *Psychology Reports* 90 (2): 451–60.

Aziz, Abdul (2004). 'Machiavellianism Scores and Self-Rated Performance of Automobile Salespersons'. *Psychology Reports* 94 (2): 464–6.

— (2005). 'Relationship between Machiavellianism Scores and Performance of Real Estate Salespersons'. *Psychology Reports* 96 (1): 235–8.

Christie, R., and F.L. Geis (1970). *Studies in Machiavellianism*. Academic Press. Activity 9.3: Machiavellianism Scale is reproduced at http://highered.mcgraw-hill.com/sites/0073381225/student_view0/chapter9/self-assessment_9_3.html.

MacNeil, Bonnie M., and Ronald R. Holden (2006). 'Psychopathy and the Detection of Faking on Self-Report Inventories of Personality'. *Personality and Individual Differences* 41 (4): 641–51.

RESEARCH SPOTLIGHT

'PRECONDITIONING AN AUDIENCE FOR MENTAL MAGIC: AN INFORMAL LOOK'

by John W. Trinkaus (published in *Perceptual and Motor Skills*, 1980)

Trinkaus tests out a simple magician's trick, then asks: 'Based on this limited inquiry, it appears that preconditioning of an audience by mentalists may well be effective. While this finding is interesting, perhaps of more interest are the implications associated with the use of this technique in situations other than mental magic. For example, is the seller's use of this method in the marketplace, to influence buyer selection, ethical?'

DIVINER INTERFERENCE

How do diviners divine? How do they achieve such dependable results? Barbara Tedlock, a distinguished professor of anthropology at the State University of New York at Buffalo, analysed the mystery. Her crystallized thoughts appear in a study published in the journal *Anthropology of Consciousness*.

In the report, Tedlock explains why other anthropologists were unwilling or unable to build what she has built – a 'theory of practice for divination'. The other anthropologists made a mistake. To them, divination is just 'the irrational weak sister of astronomy, mathematics, and medicine: a parasitic pseudoscience feeding on these more logical, rational sciences'. To Tedlock, it is more than that. The key was staring everyone in the face. 'One must', she writes, 'take what diviners say and do seriously.'

One must also be willing to study the findings of Stephen Hawking and other modern scientists. Given that scientists are now imagining gravity-bent light, faster-than-light particles, 'and other strange concepts that defy common sense reality', Tedlock says, 'why should we not approach divination with the same conceptual openness?'

Tedlock's theory applies to most kinds of divination. These include 'reading patterns of cracks in oracular bones, made from the shoulder blades of deer, sheep, pigs, and oxen, or the shells of turtles. Water-, crystal-, and star-gazing, dreaming, and the casting of lots. The taking of hallucinogenic drugs, and the contemplation of mystic spirals, amulets, labyrinths, mandalas, and thangkas. Reading natural signs such as the flight of birds or the road crossings of animals. Rod or pendulum dowsing. The Tarot, the Chinese I Ching, and the Yoruba Ifa readings, together with palmistry and geomancy. Contacting spirits to answer questions.' These practices occurred in so many places that we must assume they work, Tedlock points out.

Much of their value comes from shifting between being rational and being irrational. She gives this example: 'The anthropologist Roy Willis narrated his experience of this latter divinatory shift, which he observed when he visited Jane Ridder-Patrick, a well-known British herbalist, astrologer, and author of *A Handbook of Medical Astrology*. When he asked her to do his astrological chart, he observed that she appeared rational at first but then, about two-thirds of the way through the hour-long reading, the atmosphere changed.'

By injecting ancient, irrational practices with modern scientific analogies, Tedlock has brought anthropology to a new level of sophistication.

The field has come a long way since 1983, when Nigel Barley of the University of Oxford published his book *The Innocent Anthropologist*. Here's how Barley described his arrival in the Dowayo village of Kongle, in the Cameroons: 'By now the silence was becoming very strained and I felt it incumbent upon me to say something. I have already said that one of the joys of fieldwork is that it allows one to make use of all sorts of expressions that otherwise are never used. "Take me to your leader", I cried. This was duly translated and it was explained that the Chief was coming from his field.'

Tedlock, Barbara (2006). 'Toward a Theory of Divinatory Practice.' *Anthropology of Consciousness* 17 (2): 62–77.
Barley, Nigel (1982). *The Innocent Anthropologist*. New York: Henry Holt and Company.

MAY WE RECOMMEND

Modelling of interaction between a spatula and a human brain ☆

Kim V. Hansen *, Lars Brix, Christian F. Pedersen, Jens P. Haase, Ole V. Larsen

Department of Health Science and Technology, Aalborg University, Fredik Bajersvej 7, D1-203, Aalborg East 9220, Denmark

Abstract

This paper describes a method for surgery simulation, or more specifically a learning system of how to use a brain spatula.

'MODELLING OF INTERACTION BETWEEN A SPATULA AND A HUMAN BRAIN'
by Kim V. Hansen, Lars Brix, Christian F. Pedersen, Jens P. Haase and Ole V. Larsen (published in *Medical Image Analysis*, 2004)

The authors, who are at Aalborg University, Denmark, explain: 'The idea is to provide surgeons with a tool which can teach them the correlation between deformation and applied force.'

COPY WRONGS

Cheating, cheating was made the central theme of a special issue of the *Journal of College and Character*. Published in 2007, the issue featured two especially focused studies on the topic.

One study looked at university honour codes, which are popular in the US and are sprouting in the UK and elsewhere. Rodney Arnold, of the College of the Ozarks, together with Barbara N. Martin, Michael Jinks and Linda Bigby, of the University of Central Missouri, surveyed students at six universities in the American Midwest. The study asks 'Is there a difference in the level of academic dishonesty between colleges and universities that have incorporated an honor code system and those that have not?'

The answer: no.

But honour-coded students see things their own way. The authors report that 'students from honor code institutions perceived that the amount of academic dishonesty at their institutions was lower.'

The three honour-code campuses in the study are no mere run-of-the-mill specimens. Each has been formally recognized by the John Templeton Foundation as a 'character building' college. The Templeton Foundation's website (www.college andcharacter.org) celebrates these universities for 'shaping the ideals and standards of personal and civic responsibility'. The foundation's motto is 'How Little We Know, How Eager to Learn'.

The second study looked at that most literary form of cheating: plagiarism. Jean Liddell, a librarian at Auburn University in Alabama, and Valerie Fong, an adjunct professor at two small colleges in San Francisco, call their report 'Faculty Perceptions of Plagiarism'. Liddell and Fong learned that professors vary widely in their perception of plagiarism.

They surveyed teachers at Auburn University, asking each of them how bad the problem is. The answers spanned from never-see-it-in-my-classroom to about-three-quarters-of-my-students-do-it.

Some responses posed a statistical challenge. Liddell and Fong say: 'One faculty member responded that in his/her class, the percentage of plagiarism "cannot be estimated. I imagine it is quite [a bit] higher than any administrator would feel comfortable with".'

Nearly all the professors claimed that plagiarism was slightly worse nationwide than on their campus, slightly worse campus-wide than in their department, and slightly worse in the department than in just their own classes.

Auburn University has an official definition of what constitutes plagiarism. However, the study finds that professors' 'personal definitions are often quite different'.

One history professor put the whole matter into a fuller perspective. He or she explained that some supposedly deplorable actions are just peachy, but that plagiarism is a sin: 'There is no problem publishing something that has no original thought. In History we do that all the time. But, and this is important, you must cite all the sources from which you borrowed those thoughts.'

Arnold, Rodney, Barbara N. Martin, Michael Jinks, and Linda Bigby (2007). 'Is There a Relationship Between Honor Codes and Academic Dishonesty?'. *Journal of College and Character* 8 (2): 1–12.
Liddell, Jean, and Valerie Fong (2005). 'Faculty Perceptions of Plagiarism'. *Journal of College and Character* 6 (2).

IN BRIEF

'BEAUTY QUEENS AND BATTLING KNIGHTS: RISK TAKING AND ATTRACTIVENESS IN CHESS'

by Anna Dreber, Christer Gerdes and Patrik Gränsmark (published in *Journal of Economic Behavior and Organization*, 2013)

The authors, at Stockholm School of Economics and Stockholm University, explain: 'We explore the relationship between attractiveness and risk taking in chess. We use a large international panel dataset on high-level chess competitions which includes a control for the players' skill in chess. This data is combined with results from a survey on an online labor market where participants were asked to rate the photos of 626 expert chess players according to attractiveness. Our results suggest that male chess players choose significantly riskier strategies when playing against an attractive female opponent, even though this does not improve their performance. Women's strategies are not affected by the attractiveness of the opponent'.

CRIME DOESN'T PAY MUCH

When economists train their sights on robbers, the point traditionally is to study those who loot on the grandest, most legal scale and who are called 'financiers', and also (if there be consulting fees) to assist those persons.

Economists Barry Reilly, of the University of Sussex, and Neil Rickman and Robert Witt, of the University of Surrey, went against that tradition. They stole a hard look at the lowest class of bank robbers, the ones who physically go into bank branches, grab cash and literally leg it. Reilly, Rickman and Witt got access – exclusive and confidential access, they proudly confide – to data from the British Bankers' Association about bank robberies. In 2012 they published a study called 'Robbing Banks: Crime Does Pay – But Not Very Much'.

'Our research', they say, 'was concerned with the various factors that determine the proceeds from bank robberies; hence, we could work out (among other things) the economics (to the criminal) of attempting one, and the economics (to the banks) of trying to thwart it.'

Several factors influence both the likely gains and likely costs of a typical bank robbery. Adding more henchmen can boost the size of the haul ('Every extra member of the gang raises the expected value of the robbery proceeds by £9,033.20, on average, and other things being equal.'), but it also, of course, increases the total labour cost.

Each unsuccessful robbery attempt yields £0, and thus lowers the average gross (and net) income. The economists point out that failures, on average, incur particular expenses: 'The expected costs are the lengthy term in jail – converted into monetary terms at the robber's own conversion rate – times the probability of serving that term – that is, of being caught and convicted.'

Variable Description and Selected Summary Statistics

Description	Symbol/ units	Mean	Standard deviation	Min	Max
The amount stolen	R, pounds sterling	20,330.5	53,510.5	0	425,610.0
Bank raid successful? (dummy variable)	1 if successful, 0 if not	0.662	n/a	0	1
Number of bank staff present at time of the robbery		5.417	4.336	0	25
The number of bank raiders involved		1.637	0.971	1	6
Firearm displayed? (dummy variable)	1 if displayed, 0 if not	0.357	n/a	0	1
Fast rising screen in the banking outlet?	1 if present, 0 if not	0.118	n/a	0	1
Alarm activated during the raid?	1 if activated, 0 if not	0.854	n/a	0	1
Travel time in minutes from nearest police station to bank outlet	minutes	4.557	4.028	0	22
Location	1 if a shopping high street, 0 if not	0.670	n/a	0	1

Table 1 from 'Robbing Banks: Crime Does Pay – But Not Very Much'. The article teases: 'With access to a unique data set', the authors 'give us the low-down on the economics'.

They calculate that the average financial return on classical bank-robbing is 'a very modest £12,706.60 per person per raid', and that an industrious robber can expect, statistically, to work steadily at his trade for only about a year and a half before being caught and canned.

Reilly, Rickman, and Witt report that fewer and fewer bank robberies are being attempted, in both the UK and the US. They explain that economics – specifically, the decreasing 'expected value' of a bank robbery – accounts for the decline.

They offer small-time bank robbers an attractive alternative goal: 'In the UK, robberies from security vans are on the increase. Security vans offer more attractive pickings. Our framework provides a way of thinking about this, [and] effectively introduces a competing product into the robbers' "product space" and asks them to think about which will generate more proceeds.'

> ### Why robbing banks is a bad idea
>
> It is here that we can answer our original question of why, statistically speaking, robbing banks is a bad idea. The return on an average bank robbery is, frankly, rubbish. It is not unimaginable wealth. It is a very modest £12 706.60 per person per raid. Indeed, it is

Conclusions: 'Why robbing banks is a bad idea'

The study ends with what it says is a four-word 'lesson' – that 'successful criminals study econometrics'. It does not state, but perhaps does imply, a more valuable suggestion: successful criminals can hire econometricians.

Reilly, Barry, Neil Rickman and Robert Witt (2012). 'Robbing Banks: Crime Does Pay – But Not Very Much'. *Significance* 9 (3): 17–21.

RESEARCH SPOTLIGHT

'HONESTY AT A MOTOR VEHICLE BUREAU: AN INFORMAL LOOK'
by John W. Trinkaus (published in *Perceptual and Motor Skills*, 1980)

Trinkaus notes that he 'Assessed the veracity of people taking vision tests at a district office of a motor vehicle bureau … Results suggest that, when given an option, a sizeable percentage of people may well elect a style of behavior that is neither completely honest nor dishonest.'

TO SLEEP, PERCHANCE TO JUDGE

When a judge falls asleep in the courtroom, sometimes people are alert enough to notice – and then word gets out to the public. That's happened often enough for two doctors to decide to do something. What they did was to gather news reports about slumbering judges, write a paper about those reports, and then submit it for publication in the medical journal *Sleep*.

Dr Ronald Grunstein of the Royal Prince Alfred hospital in Sydney, Australia, and Dr Dev Banerjee of Birmingham Heartlands hospital in the UK saw their judge-filled-but-not-judgemental treatise appear in print in 2007. The headline was 'The Case of "Judge Nodd" and Other Sleeping Judges: Media, Society, and Judicial Sleepiness'.

Grunstein and Banerjee tell of fifteen cases – one in Australia, one in the UK, one in Canada, ten in that sometimes slumbering giant the US, and one at the international war crimes tribunal at The Hague. The 'Judge Nodd' story comes from Australia's New South Wales district court.

Grunstein and Banerjee write that Judge Ian Dodd 'had been reported to the State Judicial Commission for allegedly repeatedly falling asleep while listening to witness testimony and legal argument' in different cases over several years. The jury in a 2004 trial, they say, even 'commented on Judge Dodd's loud snoring'. The 'Judge Nodd' nickname arose the previous year, they let on, from jurors who had kept themselves awake to opportunities for self-amusement.

Grunstein and Banerjee explain that 'some months prior to any press reports about his sleepiness during trials, Judge Dodd … was diagnosed with obstructive sleep apnoea, and was apparently treated effectively'. The news stories led to a 'media frenzy', which led to early retirement for Judge Dodd. All of this woke Grunstein and Banerjee up to the titillating yet consequential tangle of medical, legal and moral issues.

Many judges are alert, quietly, to the undesirability of snoozing in court. Grunstein and Bannerjee point to a survey done by Professor Nancy J. King, a law professor at Vanderbilt University in Nashville, Tennessee. King asked 562 American judges about trials they had overseen recently: '69% of the judges reported cases in which jurors had fallen asleep', she writes. 'By judges' estimates, this had happened in more than 2,300 cases.'

King also asked, and learned, what those judges did to those sleepers: 'Sleeping jurors were usually awakened and offered a break, or a chance to drink water, cola, or coffee, but not reprimanded. Many other judges stated that they left it up to the lawyers to take action when jurors dozed, some noting that after all it was the lawyers who had put them to sleep.'

Grunstein, Ronald R., and Dev Banerjee (2007). 'The Case of "Judge Nodd" and Other Sleeping Judges: Media, Society, and Judicial Sleepiness'. *Sleep* 30 (5): 625–32.
King, Nancy J. (1996). 'Juror Deliquency in Criminal Trials in America, 1796–1996'. *Michigan Law Review* 94 (8): 2673–751.

MAY WE RECOMMEND

'DAYDREAMING, THOUGHT BLOCKING AND STRUDELS IN THE TASKLESS, RESTING HUMAN BRAIN'S MAGNETIC FIELDS'

by Arnold J. Mandell, Karen A. Selz, John Aven, Tom Holroyd and Richard Coppola (paper presented at the International Conference on Applications in Nonlinear Dynamics, 2010)

THE DANGERS OF SEINFELD

A medical study called 'Recurrent Laughter-induced Syncope' documents the existence of a paradoxical malady. Prior to reading the report, physicians might assume the phenomenon to be nothing more than a joke. A joke, they will learn, is just the beginning of the problem.

The two authors, Drs Athanasios Gaitatzis and Axel Petzold,

describe their case in simple, albeit technical, prose: 'A healthy 42-year-old male patient presented to the neurology clinic with a long history of faints triggered by spontaneous laughter, especially after funny jokes … There was no evidence to suggest cardiogenic causes, epilepsy, or cataplexy and a diagnosis of laughing syncope was made.' Gaitatzis is based at the SEIN-Epilepsy Institute in Heemstede, the Netherlands, and Petzold, at the UCL Institute of Neurology in London.

Sadly, the doctors withhold information that would specifically identify the jokes. 'The first episode occurred 17 years earlier while laughing', they say. 'A year later, a particularly funny joke triggered spontaneous heavy laughter followed by another brief episode of loss of consciousness.' But they do not tell the joke. Nor do they allude to its subject, structure or resolution.

They do offer something that's distantly akin to a punchline: 'He has suffered similar episodes ever since, where he would pass out after spontaneous, unrestrained, heavy laughter.'

They reveal only spare, peripheral information about the man's recurrent episodes of loss of consciousness. His swoons had been witnessed, and described, by bystanders. He would lose consciousness for just a few seconds. There were no characteristic body movements except, say the doctors, 'some mild, non-sustained twitching of his limbs seen in the last attack that lasted 15 seconds'.

The medical treatise resonates, perhaps unintentionally, with an early Monty Python sketch about a joke so funny that it was lethal ('No one could read it, and live').

Gaitatzis and Petzold searched through medical archives and discovered accounts of some fifteen non-fictional cases. The earliest appeared in a 1997 issue of the journal *Catheterization and Cardiovascular Diagnosis*, under the headline ' "Seinfeld Syncope".'

Drs Stephen Cox, Andrew Eisenhauer and Kinan Hreib at the Lahey Hitchcock Medical Center, Burlington, Massachusetts, treated a sixty-two-year-old man who had fainted on at least three occasions, each while 'watching the television show *Seinfeld*, specifically, the antics of the George Costanza character played by Jason Alexander. While laughing hysterically, the patient suffered sudden syncope with spontaneous recovery of consciousness within a minute. During one event, he fell face first into his evening meal and was rescued by his wife. The patient and his family were adamant that syncope has not resulted from any other television sitcoms or other stimuli.'

Drs Cox, Eisenhauer and Hreib conclude with happy news. Seinfeld syncope, they say, 'is curable by percutaneous stenting'.

Gaitatzis, Athanasios, and Axel Petzold (2012). 'Recurrent Laughter-induced Syncope'. *The Neurologist* 18 (4): 214–5.
Cox, Stephen V., Andrew C. Eisenhauer and Kinan Hreib (1997).' "Seinfeld Syncope"'. *Catheterization and Cardiovascular Diagnosis* 42 (2): 242.

UNHAPPY? YOU SHOULD ASK

Are the Russians as unhappy as they say they are? Ruut Veenhoven was so worried about that question that he wrote a study about it. Veerhoven did not skedaddle about the shrubbery; he titled his study 'Are the Russians as Unhappy as They Say They Are?'

Happiness is something of an obsession for Veenhoven. Based at Erasmus University Rotterdam, he is the editor of the *Journal of Happiness Studies*, where the 'unhappy Russians' study was published.

In poking at the Russian psyche, Veenhoven limited himself to considering just the Russians who were living in Russia during the 1990s. He evaluated neither the Russians of earlier eras nor the perhaps apocryphal moaning Russians abroad, who enjoy their unhappiness expatriotically. In his report, Veenhoven acknowledges all those masses of historical and wandering Russians with the simple statement: 'The Russians have a firm reputation for being an unhappy people.'

Why did Professor Veerhoven undertake this research?

Because, he says, he was skeptical about certain things he had been reading: 'Since the 1980s, several polls in Russia have included questions about happiness. The responses to these questions were quite similar. Average happiness was low in comparison to other nations and declined over time.'

Veerhoven saw a problem. 'There are doubts', he writes, 'about the validity of these self-reports.' He challenged himself to either verify or dismiss those doubts. After amassing and examining evidence, he reached a conclusion: 'It appears that the Russians are as unhappy as they say they are, and that they have good reasons to be so.'

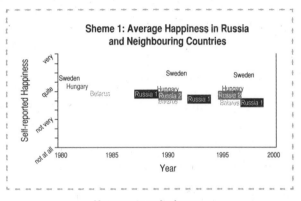

Not very to quite happy

Veenhoven built on bits and pieces of research by, among others, a professor with the delightful name 'Zapf'. Veenhoven did his writing during the year 2000, and got it into print the following year.

At about that same time, the American scholar Patricia Herlihy looked back at Russia's happy-and-unhappy national relationship with potent liquor. Herlihy titled her study '"Joy of the Rus": Rites and Rituals of Russian Drinking'. 'Joy of the Rus', she explains, is a phrase from the tenth century, credited to Prince Vladmir of Kiev in praise of booze. The joy from 'joy' would, in the coming thousand years, be part of a moody, mixed cocktail of emotions:

By the late nineteenth century ... drinking in Russian society had become something other than an unmitigated joy ... popular medicine developed aversion techniques to free the drunkard from his thirst for 'the devil's blood'. The goal was to concoct a drink so disgusting that the drunkard would react against it or be unable to stomach it. It may be that the temperance advocates, to whom we owe some of these recipes, made them especially nauseating, so as to convince readers to stay sober! The drunkard seeking to break his addiction gulped down his vodka with eels, mice, fish, the sweat of a white horse, the placenta of a black pig, vomit, snakes and worms, grease, maggots, urine, water used to wash corpses, and other disgusting things.

Vodka, especially, survived these efforts at bastardization, and in the twentieth century became the dominant alcoholic element in the Russian internal struggle over happiness.

Questions linger. Will the Russians ever find happiness? If so, is it possible that some day they will become too happy? A study by David A.F. Haaga and his colleagues, which perhaps inevitably appeared in the *Journal of Happiness Studies*, seems to anticipate this concern. The researchers, at the American University in Washington, DC, called their paper 'Are the Very Happy Too Happy?'.

In it, they nod to a study by investigators Ed Diener of the University of Illinois and Martin E.P. Seligman of the University of Pennsylvania, who concluded, rather flatly, that 'being very happy does not seem to be a malfunction.' But on the basic question of the very happy being too happy, Haaga and his fellows have not themselves taken a permanent stand. 'Future studies', they insist, 'should address this possibility.'

Veenhoven, Ruut (2001). 'Are the Russians as Unhappy as They Say They Are?'. *Journal of Happiness Studies* 2 (2): 111–36.

Herlihy, Patricia (1991). ' "Joy of the Rus": Rites and Rituals of Russian Drinking'. *Russian Review* 50 (2): 131–47.

Friedman, Elisha Tarlow, Robert M. Schwartz and David A.F. Haaga (2002). 'Are the Very Happy Too Happy?'. *Journal of Happiness Studies* 3(4): 355–72.

Diener, Ed, and Martin E.P. Seligman (2002). 'Very Happy People'. *Psychological Science* 13 (1): 81–4.

THE INHUMANITY OF PROZAC

Some have concluded that the drug fluoxetine – commonly known as Prozac – 'makes people more human'. If that conclusion is true, we can probably learn something very, very interesting from the long series of experiments in which Prozac has been shared with other members of the animal kingdom. Here are a few of the reports that document this happy scientific adventure.

Bears pacing on Prozac. 'Use of Fluoxetine for the Treatment of Stereotypical Pacing Behavior in a Captive Polar Bear' by E.M. Poulsen, V. Honeyman, P.A. Valentine and G.C. Teskey (published in the *Journal of the American Veterinary Medical Association*, 1996)

Voles socializing on Prozac. 'Effects of the Selective Serotonin Reuptake Inhibitor Fluoxetine on Social Behaviors in Male and Female Prairie Voles (*Microtus ochrogaster*)' by C. Villalba, P.A. Boyle, E.J. Caliguri and G.J. DeVries (published in *Hormones and Behavior*, 1997)

Pigeons tumbling on Prozac. 'Serotonergic Involvement in the Backward Tumbling Response of the Parlor Tumbler Pigeon' by G.N. Smith, J. Hingtgen and W. DeMyer (published in *Brain Research*, 1987)

Serotonergic involvement in the backward tumbling response of the parlor tumbler pigeon

Greg N. Smith[1], Joseph Hingtgen[2] and William DeMyer[1]

Goats ventilating on Prozac. 'Effects of Serotonin Reuptake Inhibition on Ventilatory Control in Goats' by D.R. Henderson, D.M. Konkle and G.S. Mitchell (published in *Respiration Physiology*, 1999)

Turkeys being turkeys on Prozac. 'Serotonergic Stimulation of Prolactin Release in the Young Turkey (*Meleagris gallopavo*)' by S.C. Fehrer, J.L. Silsby and M.E. El Halawani (published in *General and Comparative Endocrinology*, 1983)

Dogs dominating on Prozac. 'Use of Fluoxetine to Treat Dominance Aggression in Dogs' by N.H. Dodman, R. Donnelly, L. Shuster, F. Mertens, W. Rand and K. Miczek (published in the *Journal of the American Veterinary Medical Association*, 1996)

Cats spraying on Prozac. 'Effects of a Selective Serotonin Reuptake Inhibitor on Urine Spraying Behavior in Cats' by P.A. Pryor, B.L. Hart, K.D. Cliff and M.J. Bain (published in the *Journal of the American Veterinary Medical Association*, 2001)

Mice phagocytosing on Prozac. 'Effects of Fluoxetine on the Activity of Phagocytosis in Stressed Mice' by M. Freire-Garabal, M.J. Núñez, P. Riveiro, J. Balboa, P. López, B.G. Zamarano, E. Rodrigo and M. Rey-Méndez (published in *Life Sciences*, 2002)

Lobsters retreating on Prozac. 'Serotonin and Aggressive Motivation in Crustaceans: Altering the Decision to Retreat' by R. Huber, K. Smith, A. Delago, K. Isaksson and E.A. Kravitz (published in the *Proceedings of the National Academy of Sciences*, 1997)

Lizards changing on Prozac. 'Behavioral Changes in *Anolis carolinensis* Following Injection with Fluoxetine' by A.W. Deckel (published in *Behavioural Brain Research*, 1996)

Croakers maturing on Prozac. 'Stimulatory Effects of
Serotonin on Maturational Gonadotropin Release
in the Atlantic Croaker, *Micropogonias undulatus*'
by I.A. Khan and P. Thomas (published in *General
and Comparative Endocrinology*, 1992)

Worms moving on Prozac. 'The Effects of Cholinergic
and Serotoninergic Drugs on Motility in Vitro of
Haplometra cylindracea (Trematoda: Digenea)' by
D.M. McKay, D.W. Halton, J.M. Allen and I. Fairweather (published in *Parasitology*, 1989)

The most celebrated case may be that of Peter Fong, of Gettys-
burg College in Pennsylvania, who gave Prozac to clams. This
caused two things to happen. First, the clams began reproducing
furiously – at about ten times their normal rate. And second,
Professor Fong was awarded the 1998 Ig Nobel Prize in biology,
honouring him, his Prozac and his reproducibly happy shellfish.

Fong, Peter F., Peter T. Huminski and Lynette M. D'urso (1998). 'Induction and Potentiation of Parturition in Fingernail Clams (*Sphaerium striatinum*) by Selective Serotonin
Re-uptake Inhibitors (SSRIs)'. *Journal of Experimental Zoology* 280 (3): 260–4.

DEFENDING PAY PHONES AND PARKING SPOTS

As pay telephones disappear from our cities, with them vanish
opportunities to watch an entertaining, maddening form of
behaviour. The behaviour was documented in a study called
'Waiting for a Phone: Intrusion on Callers Leads to Territorial
Defense'. The report came out in 1989, before mobile phones
nudged public pay phones towards oblivion.

Professor R. Barry Ruback, with some of his students at
Georgia State University, performed an experiment. They began
by asking people what they would do if, while talking on a
public pay telephone, they noticed someone else waiting to
use that phone. Most people said they would hurry up and
terminate their call.

The researchers put that common belief to the test. They lurked discreetly near public telephone booths in the Atlanta area. Seeing someone engaged in a call on a pay phone, they would send a trained stooge to hover expectantly. The stooge 'simply stood behind the caller, sometimes looking at his watch and putting his hands in his pockets'. Sometimes they sent two stooges. Every stooge was 'instructed not to stare at the subject'.

In the absence of stooges, people's phone calls lasted on average about 80 seconds. When a single stooge stood nearby, people stayed on the phone longer – typically about 110 seconds. And when two stooges queued up, clearly waiting, waiting, waiting for access to the telephone, people kept using that phone much longer – averaging almost four minutes.

After varying the experiment in small ways, trying to tease out exactly what was or wasn't happening, the researchers decided they had seen a clear cause and effect – that 'people stayed longer at the phone after an intrusion, primarily because someone was waiting to use the phone'.

Even in the absence of pay phones, one can, while strolling through town, see bursts of this kind of 'territorial defense'. They happen in the street and in parking lots, wherever motorists vie for parking spaces.

Professor Ruback went behaviour-hunting in a shopping mall car park near Atlanta. In 1997, he and a colleague, Daniel Juieng, produced a report with a title that hints at more violence than the paper delivers: 'Territorial Defense in Parking Lots: Retaliation Against Waiting Drivers'. When the researchers saw someone get into a car, preparing to drive away, they measured the time until the car actually departed. They saw that, consistently, drivers took longer to leave if someone else was obviously waiting for their space.

Ruback and his minions forced the issue, sending their own drivers, in various cars, all with particular instructions. They learned that if their 'intruding' driver honked a horn, the departing driver would take an especially long time to leave. They also learned that men typically would leave more quickly if they saw that the person waiting to take their place drove a blatantly more expensive vehicle. Women, though, usually were not cowed by such things.

Ruback, R. Barry, Karen D. Pape and Philip Doriot (1989). 'Waiting for a Phone: Intrusion on Callers Leads to Territorial Defense'. *Social Psychology Quarterly* 52 (3): 232–41.
Ruback, R. Barry, and Daniel Juieng (1997). 'Territorial Defense in Parking Lots: Retaliation Against Waiting Drivers'. *Journal of Applied Social Psychology* 27 (9): 821–34.

RED: BULL

Bulls care little about the redness of a matador's cape. Psychologists have been pretty sure about that since 1923, when George M. Stratton of the University of California published a study called 'The Color Red, and the Anger of Cattle'.

'It is probable', Stratton opined, 'that this popular belief arises from the fact that cattle, and particularly bulls, have attacked persons displaying red, when the cause of the attack lay in the behavior of the person, in his strangeness, or in other factors apart from the color itself. The human knowledge that red is the color of blood, and that blood is, or seemingly should be, exciting, doubtless has added its own support to this fallacy.'

Professor Stratton, aided by a Miss Morrison and a Mr Blodgett, conducted an experiment on several small herds of cattle – forty head altogether, a mixture of bulls and bullocks (bullocks are castrated bulls) and cows and calves, including some who were accustomed to wandering the range and others who lived in barns.

The researchers obtained white, black, red and green strips of cloth, each measuring two by six feet. These they attached 'endwise to a line stretched high enough to let the animals go easily under it; from this line the colors hung their 6 feet of length free of the ground, well-separated, and ready to flutter in the breeze.'

The cattle showed indifference to the banners, except sometimes when a breeze made the cloth flutter. Males and females reacted the same way, as did 'tame' and 'wild' animals. Red did nothing for them.

Farmers seem to have already suspected this. Stratton surveyed some. He reports that 'Of 66 such persons who have favored me with their careful replies, I find that 38 believe that red never excites cattle to anger; 15 believe that red usually does not excite them to anger, although exceptionally it may; 8 believe that it usually so excites, though exceptionally it may not; and 3 believe that it always so excites.'

One of those three dissenters described her views, well, colourfully: 'A lively little Jersey cow whom I had known all her six years of life, chased me through a barbed wire fence when I was wearing a red dress and sweater, and never did so before or after. I changed to a dull gray, and reentered the corral, and she paid no attention to me, and let me feed and water her as usual. Also a Durham bull whom I had raised from a calf, and was a perfect family pet, chased me till I fell from sight through some brush when I was wearing the same outfit of crimson.'

More typical, though, was the farmer who told Stratton: 'In referring to the saying, "Like waving a red rag before a bull, I have found that to wave anything before a bull is dangerous business." '

Stratton, George M. (1923). 'The Color Red, and the Anger of Cattle'. *Psychological Review* 30 (4): 321–5.

IN BRIEF

'AN EIFFEL PENETRATING HEAD INJURY'
by M. George and J. Round (published in *Archives of Disease in Childhood*, 2006)

The authors, at St George's Hospital Medical School in London, report that: 'A 3 year old boy presented to our accident and emergency department with an obvious penetrating head injury. He had tripped and fallen onto a metal model of the Eiffel Tower which then became rigidly lodged into his skull.'

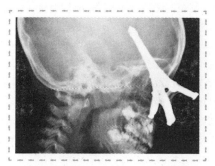

X-ray of a three-year-old boy's head with souvenir Eiffel Tower

INSULTING IN FRENCH

Prior to 2008 no one knew, at all precisely, the pain people suffer when they gaze at an ugly painting – relative to what they'd feel if they were looking at a pretty picture – while a stranger shoots them in the back of the hand with a powerful laser beam. Now something is known about the subject. The knowledge is preserved in a study called 'Aesthetic Value of Paintings Affects Pain Thresholds'.

The study's authors, Marina de Tommaso, Michele Sardaro and Paolo Livrea at the University of Bari, Italy, had twelve people each identify paintings as beautiful or ugly, then stare at some of each kind while a laser heated into the dorsal surface of their hand. Each volunteer, after each viewing, rated the pain on a scale of 0 to 100. The hurt was a little worse when they looked at ugly art, they said, mostly.

This manner of inflicting pain – applying a carefully aimed column of light amplified by stimulated emission of radiation (that's the phrase, more or less, that gives us the cool, five-letter word 'laser'), is not the only possible way. In analysing how people respond to pain, researchers have dabbled or experimented with different methods of causing that pain. Harold Hillman, of the University of Surrey, published a paper in 1993 called 'The Possible Pain Experienced During Execution by Different Methods'. His key observation was: 'It is difficult to know how much pain the person being executed feels, or for how long, because many of the signs of pain are obscured by the procedure.' In contrast, Richard Stephens of Keele University and his colleagues plunged people's hands into ice-cold water in their 2009 study 'Swearing as a Response to Pain', and again in the 2011 follow-up, 'Swearing as a Response to Pain – Effect of Daily Swearing Frequency'.

Swearing is risky for researchers. They must beware of the foreign-language discount. Several psychologists have found that swearing in one's native language dredges up deeper, more hellacious emotion than swearing in a 'foreign' language. Among the more subdued descriptions of this quirk, one finds a paper called 'The Emotional Force of Swearwords and Taboo Words in the Speech of Multilinguals' by Jean-Marc Dewaele of Birkbeck, University of London. It was published in 2004. Six years later Dewaele produced a study with a more colourful title, one that risks inflicting pain on any journal or newspaper editor who considers permitting a writer to mention it: 'Christ Fucking Shit Merde'.

The authors of the basic research on people's reactions to being shot with a laser beam while watching ugly or pretty paintings also continued researching and writing. De Tommaso, Sardaro and Livrea repeated their original experiment, this time on people who were already prone to heady suffering. The result: a 2009 paper called 'Effects of Affective Pictures on Pain Sensitivity and Cortical Responses Induced by Laser Stimuli in Healthy Subjects and Migraine Patients'.

Tommaso, Marina de, Michele Sardaro and Paolo Livrea (2008). 'Aesthetic Value of Paintings Affects Pain Thresholds'. *Consciousness and Cognition* 17 (4): 1152–62.
Hillman, Harold (1993). 'The Possible Pain Experienced during Execution by Different Methods'. *Perception* 22 (6): 745–53.

Stephens, Richard, John Atkins and Andrew Kingston (2009). 'Swearing as a Response to Pain'. *Neuroreport* 20 (12): 1056–60.

Stephens, Richard, and Claudia Umland (2011). 'Swearing as a Response to Pain – Effect of Daily Swearing Frequency'. *Journal of Pain* 12 (12): 1274–81.

Harris, Catherine L., and Jean Berko Gleason (2003). 'Taboo Words and Reprimands Elicit Greater Autonomic Reactivity in a First Language than in a Second Language'. *Applied Psycholinguistics* 24 (4): 561–79.

Dewaele, Jean-Marc (2004). 'The Emotional Force of Swearwords and Taboo Words in the Speech of Multilinguals'. *Journal of Multilingual and Multicultural Development* 25 (2&3): 204–22.

Dewaele, Jean-Marc (2010). 'Christ Fucking Shit Merde! Language Preferences for Swearing among Maximally Proficient Multilinguals'. *Sociolinguistic Studies* 4 (3): 595–614.

Tommaso, Marina de, Rita Calabrese, Eleonora Vecchio, Francesco Vito De Vito, Giulio Lancioni and Paolo Livrea (2009). 'Effects of Affective Pictures on Pain Sensitivity and Cortical Responses Induced by Laser Stimuli in Healthy Subjects and Migraine Patients'. *International Journal of Psychophysiology* 74 (2): 139–48.

MAY WE RECOMMEND

'DELUSIONS OF HALITOSIS'

by R.L. Goldberg, P.A. Buongiorno and R.I. Henkin (published in *Psychosomatics*, 1985)

Medical Data for Six Patients with Delusions of Halitosis

Chief complaint	Duration (years)	Precipitating event	Family history of a similar complaint	Medical history
Bad taste / breath	19	None	Parent / siblings	Sinusitis
Bad taste / breath	3	None	Parent	Sinusitis
Bad taste / breath	10	Separation from spouse	Parent	Allergies
Bad taste / breath	20	Menstruation	Parent	Sinusitis
Bad taste / breath	24	None	None	Allergies
Bad taste / breath	3	Loss of mother and son	None	Sinusitis

THREE

EARS, TONGUES, NOSES AND ALL THAT

MAY WE RECOMMEND

'POST-NASAL DRIP SYNDROME: A SYMPTOM TO BE SNIFFED AT?'
by Alyn H. Morice (published in *Pulmonary Pharmacology and Therapeutics*, 2004)

Some of what's in this chapter: Old men and their big ears • Look at everything upside-down • A bright thumb in the eye of the paparazzo • The Hapsburg eye, the Coburg nose • Print, lip • Experimental ant gulp • Pet-mounted noise alarm • Music's eyebrow • Some swallow it hot • Christian smells, holy smells • The afterlife of aristocrats' body parts • Chalk dust up the nose • Sit, hand; Bite, tongue • Monkey flossing • Cured pork up the nose

ALL EARS RESEARCH REVIEW

Old men have big ears is the consensus of several medical studies on the question. The most celebrated work focused exclusively on men, according with British male doctordom's smug tradition of showing interest mainly in themselves. But in Japan and in Germany, wide-ranging investigations broke through the patriarchal hegemony. The newer studies made plain, for anyone who cared to know, the long-untold half of the story: old women have big ears too.

The British action played out in a characteristic location: the pages of the *British Medical Journal*, where any body part is always of interest.

In 1993, Dr James A. Heathcote, a general practitioner in Bromley, 'set out to answer the question "As you get older do your ears get bigger?" ' Heathcote and three colleagues examined the ears of 206 men of various ages, then presented his findings in a monograph called 'Why Do Old Men Have Big Ears?' They report: 'A chance observation – that older people have bigger ears – was at first controversial but has been shown to be true.'

Using the pronoun 'we' in a manner that excludes half of the population, Dr Heathcote wrote: 'As we get older our ears get bigger (on average by 0.22 mm a year).'

The biggest oddity, ears aside, comes at the end. Almost in defiance of its title, the paper mutters: 'Why ears should get bigger when the rest of the body stops growing is not answered by this research.'

Outside Britain, ear-growth data-gatherers took the bother to also look at man's counterpart: woman.

In Japan, primary care physicians Yasuhiro Asai, Manabu Yoshimura, Naoki Nago and Takashi Yamada measured the ears and height of four hundred adult patients – of both sexes – who visited their clinics. The team's 1996 report called 'Correlation of Ear Length with Age in Japan', also appeared in the sometimes-seemingly 'we're-all-ears' *British Medical Journal*. The doctors claim two discoveries:

1) 'Ear length correlates significantly with age, as Heathcote showed, in Japanese people'
2) 'ear length corrected for height shows [even] greater correlation with age'

A decade later, Carsten Niemitz, Maike Nibbrig and Vanessa Zacher at Freie Universität Berlin, Germany, took a bisexual look at lots of ears. They examined data from a thesis by a scientist named Montacer-Kuhssary, published at the university in 1959. Montacer-Kuhssary's data were of a rare kind: photographs of 1,448 ears from newborn children, older children and adults up to and including ninety-two-year-olds.

For each ear, the team made fifteen different measurements. They confirmed, they say, that ears never really stop growing throughout a person's lifetime.

But the big surprise came from comparing women and men: 'In all parameters where post adult growth was observed, female ears showed a lesser increase than those of men.' Old men have bigger ears than old women.

Montacer-Kuhssary, by the way, noted back in 1959 that people's noses, too, usually grow throughout their lifetimes. But in the race toward biggerness, said Montacer-Kuhssary, ears outpace noses.

Heathcote, James A. (1995). 'Why Do Old Men Have Big Ears?'. *British Medical Journal* 311 (23 December): 1668.
Asai, Yasuhiro, Manabu Yoshimura, Naoki Nago and Takashi Yamada (1996). 'Correlation of Ear Length with Age in Japan'. *British Medical Journal* 312 (2 March): 582.
Niemitz, Carsten, Maike Nibbrig and Vanessa Zacher (2007). 'Human Ears Grow Throughout the Entire Lifetime According to Complicated and Sexually Dimorphic Patterns: Conclusions from a Cross-sectional Analysis'. *Anthopologischer Anzeiger* 65 (4): 391-413.

IN BRIEF

'GLUE EAR AND GROMMETS'
by D. Isaacs (published in the *British Medical Journal*, 1992)

Glue ear and grommets

Non-suppurative otitis media was described by Wathen in 1755[1] and again by Politzer[2] in 1867. Today it is the most common cause of loss of hearing in children.[3][4] The condition —known also as "glue ear," exudative, secretory, or catarrhal otitis media—is characterised by the collection of sterile fluid of varying consistencies in the middle ear. It is probably

AN UPRIGHT AUSTRIAN VISION

In the middle of the twentieth century, an Austrian professor turned a man's eyesight exactly upside-down. After a short time, the man took this completely in his stride.

Professor Theodor Erismann, of the University of Innsbruck, devised the experiment, performing it upon his assistant and student, Ivo Kohler. Kohler later wrote about it. The two of them made a documentary film in 1950, 'The Reversing Glasses and the Upright Vision'. (You can watch it at http://www.youtube.com/watch?v=zlHYcN789N4.)

The professor made Kohler wear a pair of hand-engineered goggles. Inside those goggles, specially arranged mirrors flipped the light that would reach Kohler's eyes, top becoming bottom, and bottom top.

At first, Kohler stumbled wildly when trying to grasp an object held out to him, navigate around a chair or walk down stairs.

In a simple fencing game with sticks, Kohler would raise his stick high when attacked low, and low in response to a high stab.

Holding a teacup out to be filled, he would turn the cup upside down the instant he saw the water apparently pouring upwards. The sight of smoke rising from a match, or a helium balloon bobbing on a string, could trigger an instant change in his sense of which direction was up, and which down.

But over the following week, Kohler found himself adapting, in fits and starts, then more consistently, to such sights.

After ten days, he had grown so accustomed to the invariably upside-down world that, paradoxically and happily, everything seemed to him normal, rightside-up. Kohler could do everyday activities in public perfectly well: walk along a crowded sidewalk, even ride a bicycle. Passers-by on the street did ogle the man, though, because his eyewear looked, from the outside, unfashionable.

Erismann and Kohler did further experiments. So did other scientists. Their impression is that many, perhaps most, maybe just about all, people are able to make these kinds of adjustment. Images reach the eye in some peculiar fashion, and if that peculiar fashion is consistent, a person's visual system eventually, somehow, adjusts to interpret it – to perceive it, to see it – as being no different from normal. Kohler writes that, 'after several weeks of wearing goggles that transposed right and left', one person 'became so at home in his reversed world that he was able to drive a motorcycle through Innsbruck while wearing the goggles'.

This may strike you as extremely unusual. But the basic ability – to adapt to visions seen topsy-turvy or backwards – is something you have almost certainly witnessed. Many people develop the ability to read documents that are upside down. Many teachers, especially, treasure this as a semi-secret skill they've picked up without having worked at it.

This automatic, almost-effortless adaptation to visual weirdness is one of many bizarre things that brains do that scientists simply do not understand. Were we not talking about the brain, it would be appropriate to say that these behaviours, these abilities, are so weird that they are 'unthinkable'.

Kohler, Ivo (1962). 'Experiments with Googles'. *Scientific American* 206: 62–72.

FLASHY WAYS TO FIGHT OFF PAPARAZZI

A new invention aims to foil paparazzi who try to photograph people who do not wish to be photographed. Wilbert Leon Smith Jr and Keelo Lamance Jackson of California obtained a patent in 2012 for what they call 'Inhibiting Unwanted Photography and Video Recording'. Their invention builds on a simple idea patented in 2005 by Jeremy and Joseph Caulfield from Arizona.

The Caulfields equipped celebs with a flashgun that fires automatically the instant another flashgun fires nearby. Smith and Jackson's device goes that bit better: it's a rotating, swivelling, oscillating device that can emit multiple strobe lights and other light beams for as long as the celebrity deems necessary.

The device has uses beyond deterring pesky paparazzi. As Smith and Jackson explain, it can also protect our own spy agencies against nosy foreign bad guys: 'A surveillance camera detects a covert government operative with access to photographic equipment outside of the government building. Once the covert government operative is detected, the image distortion apparatus subsequently emits a plurality of deterrents in the direction of the photographic equipment to distort images captured therein.'

Inhibitions demonstrated: A celebrity fights off a paparazzo (left) and a soldier shows off a helmet-mounted version of the apparatus (inset)

Paparazzi and spies are but a tiny segment of the population. A decade earlier, Maurizio Pilu, a researcher at Hewlett-Packard in Bristol, took aim at the more general problem. Whenever people set foot in a public place, they can be photographed by strangers who have tiny (or, for that matter, big) cameras. This can happen countless times a day without anyone realizing it. Stroll down a street, and your image may have been captured in images by hundreds of people who were intent on photographing fire hydrants, cats or some civic official who waved at the populace while riding a bicycle.

Pilu's method could prevent these unwanted captures. Anyone who wants privacy would carry on their person a special signalling device that transmits an electromagnetic message that indicates 'Do not photograph me'. The scheme requires, perhaps quixotically, that every camera – every camera, owned by anyone – has a special circuitry built into it to receive such signals and alter the camera's behaviour accordingly.

Pilu no longer works at Hewlett-Packard. But he still keeps an eye on the problem in his new job as lead technologist for digital in the Innovation Programme Directorate at the UK government's executive innovation agency, the Technology Strategy Board.

Pilu has clocked the arrival of Google Glass – the expensive, not readily available video camera/computer/transmitter/receiver that allows its owner constantly to gather video imagery of whatever happens to be in front of them, wherever they go. On 8 March 2013, in a message sent via Twitter, Pilu (@Maurizio_Pilu) warned the world that Google Glass is just the beginning and that cheaper alternatives to it are coming.

Caufield, Joseph A. (2005). 'Apparatus and method for preventing a picture from being taken by flash photography'. US patent no. 6,937,163, 30 August.
Jackson, Keelo Lamance, and Leon Smith Wilbert Jr (2012). 'Inhibiting unwanted photography and video recording'. US patent no. 8,157,396, 17 April.
Pilu, Maurizio (2004). 'Image capture method, device and system'. US patent no. 7,653,259, 14 October.

IN BRIEF

'THE HISTORY OF THE EVIL EYE AND ITS INFLUENCE ON OPHTHALMOLOGY, MEDICINE AND SOCIAL CUSTOMS'
by George H. Bohigian (published in *Documenta Ophthalmologica*, 1997)

FEATURES, PREDICTABLY

Researchers in one field do not always pick up on good suggestions from those outside their specialty. Take, for example, the case of the Hapsburg lip.

'I do not propose to deal with one of the most famous inherited features, the "Hapsburg Lip" … because it could almost be described as a medical condition, about which I am not qualified to speak. However, I feel sure that the "Hanoverian Eye", the "Coburg Nose" and the "Danish Neck" will prove equally fascinating.' So said Frances Dimond, curator of the Royal Photographic Collection, in a lecture that that was published in 1994 in *The Genealogists' Magazine*. In the years since, however, the biomedical research community has displayed a collective lack of curiosity about the Hapsburg, Hanoverian and Coburg lip, eye and nose.

Partly this is because certain aspects of these phenomena *are* well understood. Dimond pointed this out when she said: 'The true enthusiast for the Coburg nose was, however, the Queen and Prince Albert's cousin, Prince Augustus of Saxe-Coburg-Gotha, who, possessing a fine nose himself, married a French princess, Clementine, who was similarly endowed, with predictable results.'

One has to search the medical literature back to 1988 to find more than a cursory mention of these matters. That was the year that E.M. Thompson and R.M. Winter of the Institute of Child Health in London published their report 'Another Family With the "Habsburg Jaw" '. Their report does not stint on detail: 'We report a three generation family with similar facial characteristics to those of the Royal Habsburgs, including mandibular prognathism, thickened lower lip, prominent, often misshapen nose, flat malar areas, and mildly everted lower eyelids.'

A year earlier, W. Neuhauser had gone into even more depth, in his study entitled 'Example of Potentiation of Genetic Traits Due to Inbreeding: The Habsburg Chin, Burgundian Lip, Spanish

Insanity'. But because it was published in the German journal *Zahnärztliche Mitteilungen*, Neuhauser's work is little-read in English-speaking countries.

A neurologist of my acquaintance recently moved to east Texas, where he discovered a most unexpected source of research material. He reports that, thanks to many generations of inbreeding, the region is full of genetically based neurological phenomena that he had seen only in medical books. What he had thought to be rare curiosities turn out to be commonplace in Texas. Inbreeding can, indeed, produce offspring with unusual characteristics.

The royal families of Europe and the hoi polloi of east Texas are both there, quietly waiting for scientists to come study and make sense of them.

Dimond, Frances (1994). 'Face Values: Inherited Features'. *The Genealogists' Magazine* 24 (10): 893–908.
Thompson, E.M., and R.M. Winter (1988). 'Another Family with the "Habsburg Jaw"'. *Journal of Medical Genetics* 25 (12): 838–42.
Neuhauser, W. (1977). 'Example of Potentiation of Genetic Traits Due to Inbreeding: The Habsburg Chin, Burgundian Lip, Spanish Insanity'. *Zahnärztliche Mitteilungen* 67 (15): 914–6.

IN BRIEF

'BROKEN HAND OR BROKEN NOSE: A CASE REPORT'
by B.P. Shravat and S.N. Harrop (published in the *Journal of Accident and Emergency Medicine*, 1995)

ON EVERYONE'S LIPS

They are on everyone's lips always, and sometimes on a shred of evidence in a murder trial, and occasionally in the title of a scientific report (as in the recently published 'Morphologic Patterns of Lip Prints in a Portuguese Population: A Preliminary Analysis').

Lip prints – lip patterns – have become the subject of formal study. That formal study has a formal name: cheiloscopy.

Basic questions still nag at cheiloscopists.

The Portuguese population lip print patterns paper, written by Virgínia Costa and Inês Caldas of the Universidad do Porto, appears in the *Journal of Forensic Sciences*. Costa and Caldas show how scientists have worked hard to classify the universe of lip patterns into a set of standard categories. They slightly lament the existence of competing standards, the field being too new for its experts to settle on a single taxonomy.

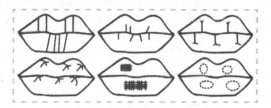

'Suzuki and Tsushihashi's classification' scheme for prints

They also worry at the how-do-the-lips-change-after-death conundrum. Criminal investigators find themselves haunted by a scarcity problem with both before-and-after patterns. 'Very few' corpses arrive with a companion set of pre- and postmortem lip prints, Costa and Caldes say, 'which obviously impairs a comparative study'.

Does each of the seven billion or so people on earth (and each of their ancestors) have a unique set of lip prints? Jerzy Kasprzak of the Military Police School in Minisk Mazowiecki, Poland, addressed that in 1990, in an article called 'Possibilities of Cheiloscopy', in the journal *Forensic Science International*. 'Cheiloscopy deals with the examination of systems of furrows on the red part of human lips', he specified, then explaining that Yasuo Tsuchihashi and Tazuo Suzuki at Tokyo University had examined the lips of 1,364 persons. Thanks to those 1,364 sets of lips, Kasprzak said, Tsuchihashi and Suzuki 'established that the arrangement of lines on the red part of human lips is individual and unique for each human being'.

Materials

Lipstick smears	Fingerprint powders	Reagents
Long-lasting lipstick	Fingerprint red powder (Dragon's Red)	Ninhydrin
Unperfumed tissue paper	Fingerprint black powder	Acetone
White cotton fabric	Silver metallic powder	Sudan III
Bottles (dark storage, sprayer, plastic)		Oil Red O
Brushes		Sudan black
Fume hood and fuming chamber		Ethanol
Glassware		Distilled water
Heater		
Humidity chamber		
Stirring device		

According to the paper, 'After applying the lipstick on ten volunteers and waiting the recommended five minutes for the lipstick to fix, lip impressions were made on tissue paper and white cotton fabric using sustained pressure for three seconds. The samples were exposed to ambient conditions.'

Do women's lips have identifiably different patterns from men's? In 2009, a team at Subharti Dental College in Meerut, India, attacked the question, using 'lipstick, bond paper, cellophane tape, a brush for applying the lipstick, and a magnifying lens'. Their resulting treatise, 'Cheiloscopy: The Study of Lip Prints in Sex Identification', reports success in identifying the gender of eighteen of twenty women and seventeen of twenty men.

Lips being often associated with romance, cheiloscopy smacks occasionally of glamour. Ana Castelló, Mercedes Alvarezu, Fernando Verdú of the University of Valancia, Spain, noticed that advanced developments in the fashion industry had forced crime-fighters to come up with their own, countervailing technological leaps. In 2002, they and colleague Marcos Miquel published a study called 'Long-lasting Lipsticks and Latent Prints', in which they complained that 'the cosmetics industry has developed long-lasting lipsticks that often do not leave visible prints'.

**Evidence: 'Latent lip print on cotton fabric developed
using Oil Red O (powder) after 30 days'**

The Valencia team experimented with chemicals – especially a dye called Nile red – that helped reveal nearly invisible prints left by such lipsticks. That led them, three years later, to publish one of the most romantically titled reports in the history of forensics: 'Luminous Lip-prints as Criminal Evidence'.

Costa, Virgínia A., and Inês M. Caldas (2012). 'Morphologic Patterns of Lip Prints in a Portuguese Population: A Preliminary Analysis'. *Journal of Forensic Sciences* 57 (5): 1318–22.

Kasprzak, Jerzy (1990). 'Possibilities of Cheiloscopy'. *Forensic Science International* 46 (1/2): 145–51.

Sharma, Preeti, Susmita Saxena and Vanita Rathod (2009). 'Cheiloscopy: The Study of Lip Prints in Sex Identification'. *Journal of Forensic Dental Sciences* 1 (1): 24–7.

Castelló, Ana, Mercedes Alvarez, Marcos Miquel and Fernando Verdú (2002). 'Long-lasting Lipsticks and Latent Prints'. *Forensic Science Communications* 4 (2): n.p.

Castelló, Ana, Mercedes Alvarez-Seguí and Fernando Verdú (2005). 'Luminous Lip-prints as Criminal Evidence'. *Forensic Science International* 155 (2/3): 185–7.

A SNACK WITH BITE

> *There once was a man who swallowed some ants.*
> *'Twas done with intent, not merely from chance.*
> *The ants were alive,*
> *But did not survive.*
> *The research was done without government grants.*

The man in verse was and is Volker Sommer, professor of evolutionary anthropology at University College London. He and his colleagues Oliver Allon and Alejandra Pascual-Garrido travelled to Nigeria's Gashaka Gumti national park. There, chimpanzees and army ants and sticks are plentiful – the former use the latter to dip into nests for presumably delicious helpings of fresh, lively army ants of the species *Dorylus rubellus*. As the scientists describe it: 'Army ants respond to predatory chimpanzees in a particular way by streaming to the surface to defend their colony through painful bites. In response, chimpanzees typically harvest army ants with stick tools, thereby minimizing the bites they receive.'

The team craved more knowledge about this chimp/ant give-and-take. So they 'mimicked the predatory behaviour of tool-using chimpanzees at army ant nests to study the insects' response'.

How? They used discarded chimp-manufactured 'dipping wands'.

Some things were fairly easy to measure: how fast ants run up a dipping wand; how many ants one can expect to harvest in a single dip; the typical weight of those ants.

The most demanding part of the research is described in the report section headlined: 'Ant remains in chimpanzee faeces and self-experiments'.

The goal was to estimate the numerical relationship between (a) ants that go into a chimp and (b) discernible ant parts that come out.

Anyone can wander through the park, collect chimp faeces and try to count its ant content. The pure measurement aspect is straightforward, if one utilizes simple accounting techniques. Here's a technique mentioned in the report: 'Counts of broken ant heads (usually only those of large workers could still be identified as such) were divided by two and added to the number of complete heads.'

Number of ingested and excreted ants

The correlation between ingested ants and those detectable in faeces was assessed through self-experiments, in which a human volunteer chewed and swallowed batches of 100 large *Dorylus* workers. The number of detectable heads (H) and other fragments (O) and intervals (in hours) between ingestion and excretion was as follows: experiment 1 (11 H/14 O/17 h; 10 H/2 O/23 h); experiment 2 (11 H/9 O/15 h; 7 H/6 O/27 h; 23 H/18 O/39 h; 3 H/4 O/51 h; 5 H/1 O/67 h); experiment 3 (8 H/2 O/16 h; 31 H/10 O/38 h; 2 H/1 O/45 h; 0 H/0 O/68 h).

Detail of methods in Allon, Pascual-Garrido and Sommer (2012)

But measuring how many ants go into a chimp requires an expensively high degree of controlled, intimate access to the animal. Sommer, Allon and Pascual-Garrido suggest that some day this might be done: 'Controlled experiments in which captive chimpanzees are fed known numbers of army ants followed by faecal inspections could clarify this issue further.'

In the meantime, Sommer offered himself as a substitute for a chimp: 'Correlations between numbers of ingested ants and remains detectable in faeces were assessed via [three] self-experiments by VS. For each of these, 100 large Dorylus workers were immobilized in whisky, ingested, masticated with 12 chewing motions and swallowed. Remains detectable in subsequent excreta were counted.'

Sommer produced some interesting results. The study lists summary data for the 'number of detectable heads'; the number of 'other fragments'; and the 'interval (in hours) between ingestion and excretion'. Calculations suggest that one can 'assume that 10.1% of ingested insects are detectable in excreta produced during the subsequent three days'.

Allon, Oliver, Alejandra Pascual-Garrido and Volker Sommer (2012). 'Army Ant Defensive Behaviour and Chimpanzee Predation Success: Field Experiments in Nigeria'. *Journal of Zoology* 288 (4): 237–44.

AN IMPROBABLE INNOVATION

'AUTOMATED SURVEILLANCE MONITOR OF NON-HUMANS IN REAL TIME'

a/k/a a pet-mounted noise-sensing alarm, by David Shemesh and Dan Forman (US patent no. 6,782,847, granted 2004)

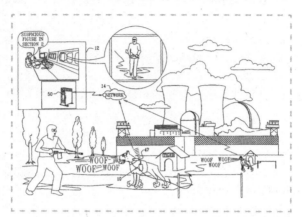

Fig 18. from US patent no. 6,782,847

THE HIGHS AND LOWS OF THE MUSICAL EYEBROW

When singers sing high notes, their eyebrows go higher than when they sing low notes. While that may not be an absolute physiological rule, a team of Danish and American researchers discovered that it happens pretty consistently. They lay out the evidence, and explain what it may mean, in a study called 'Facial Expression and Vocal Pitch Height: Evidence of an Intermodal Association', published in the journal *Empirical Musicology Review*.

When scientists tackle a new question, they begin with the knowledge that finding the answer – if there is an answer – might entail lengthy, slogging effort. Some questions take years to settle. Some take decades. The eyebrow/high-note evidence comes from 'an experiment lasting less than one minute'.

Sofia Dahl, at Aalborg University Copenhagen, and David Huron and Randolph Johnson, at Ohio State University, ran their experiment forty-four times, each with a different volunteer. They asked each person to sing, but intentionally did not solicit any information as to whether anyone had musical training.

The one thing each person did have was an ice-cream bar. The study says that each volunteer received one, as inducement to sing, and that it was free.

The experiment was simple, as well as quick. To prompt each volunteer to sing, Dahl, Huron or Johnson used this script: 'I want you to sing a comfortable pitch and sustain it while we take your picture. Sing whatever vowel you like. Now hold the note ... [Take picture]. Now I want you to sing a higher [lower] pitch – the highest [lowest] pitch you can. Good. Now hold it. [Take picture]'.

The research team then showed the photographs to judges who had not been present during the vocalization. The judging yielded results that were, on the face of them, stark. The report says the 'independent judges selected the high-pitch faces as more friendly than the low-pitch faces. When photographs were cropped to show only the eye region, judges still rated the high-pitch faces friendlier than the low-pitch faces.'

Building on a mound of earlier research that is documented in the scientific literature, Dahl, Huron and Johnson assessed the brow/pitch behaviour they saw and heard. Their data, they conclude, is 'consistent with the role of eyebrows in signaling aggression and appeasement'.

The prior work by others gathered evidence about the role or roles eyebrows play in expressing human emotion. This new study mentions a discovery in 1978 that 'raised or arched eyebrows are indicative of appeasement or friendliness', a 1979 finding that 'when angry, the eyebrows are lowered, resulting in a more pronounced brow ridge', and a 1981 treatise explaining that in many cultures lowered eyebrows are 'interpreted as displays of greater dominance or aggression'.

Having, this once, focused on a notably fleeting musical phenomenon – the brow-arched single note – Johnson eventually moved to a new institution and towards an opposite extreme. Ensconced at Oklahoma Baptist University, he presented a paper called 'The Fullness of God's Time in Brahms's Requiem'.

Huron, David, Sofia Dahl and Randolph Johnson (2009). 'Facial Expression and Vocal Pitch Height: Evidence of an Intermodal Association'. *Empirical Musicology Review* 4 (3): 93–100.
Johnson, R. (2011). 'The Fullness of God's Time in Brahms's Requiem'. Paper presented at the Forum on Music and Christian Scholarship, Wheaton College, Wheaton, Illinois, 19 March.

HEAT OVER HOTNESS

To drink really hot coffee (or hot tea) is to swallow a paradox of pleasure and pain. Hye-Seong Lee, Earl Carstens, and Michael O'Mahony, at the University of California, Davis, solved the puzzle, more or less. They explain it in a study that, for the sake of clarity and directness, they call: 'Drinking Hot Coffee: Why Doesn't It Burn the Mouth?'

Industries acquire standards. The various industries that make and serve hot beverages have acquired standard notions of how hot people expect a good cuppa to be.

ABSTRACT

Coffee served at temperatures recommended by the hospitality and food literatures for brewing and holding are above thermal pain and damage thresholds. Yet, consumers do not report pain or damage on drinking coffee at such temperatures. To investigate this discrepancy, the temperature of hot coffee before and during sipping was investigated for 18 subjects. Coffee temperature

'Drinking Hot Coffee Why Doesn't It Burn the Mouth?': Abstract

Meticulous researchers showed that these industries think that people like their coffee to be really, really hot – between 80 and 85 degrees Celsius (175 and 185 degrees Fahrenheit). An authoritative source for this kind of information is a paper entitled 'Consumer Preferred Hot Beverage Temperatures', published in 1999 by Carl P. Borchgrevink and John M. Tarras of Michigan State University and Alex M. Susskind of Cornell University in New York. Other studies, by other scholars, hint that the industry has a little problem with precision, and that hot beverages are often professionally served a few degrees hotter or cooler than that ideal.

No matter. Several other experimenters discovered that anything even nearly that hot, if placed in the mouth, tends to hurt. And it hurts in both senses: pain and damage.

Lee, Carstens and O'Mahony describe the pain studies in some detail. They speak of a Dr Yamada, who 'mapped the mouth for spots sensitive to pain, using Miller's dental broach', and of a Dr Svensson, who 'mapped pain thresholds in the mouth using argon laser stimulation'. They say that for thermal pain, a Dr Green 'measured mean thresholds in the dorsal surface of the tongue (47.8 degrees C [118.04F]) and the inner wall of the lower lip (47.5 degrees C [117.5F]) with an ascending series method'. They also describe the damage studies, which can be boiled down to a single word: *burns*.

'Thus', they write, 'it would seem that for a substantial number of people, the preferred temperatures for drinking were not only above reported pain thresholds, but also above possible damage thresholds in the mouth.'

Lee, Carstens and O'Mahony answered the 'why don't people burn their mouths?' question by sticking electronic sensors inside people's mouths. They used thermocouples to measure the temperatures at four locations inside the mouths of eighteen coffee-drinkers while those coffee-drinkers drank hot coffee. One thermocouple was placed on the anterior dorsal surface of the tongue, near the tip. The others were situated to measure the bolus – the roiling slurp – of coffee as it passed through the mouth.

After all the measuring and analysing, they concluded that, probably, 'during drinking, the bolus of hot coffee is not held in the mouth long enough to heat the epithelial surfaces sufficiently to cause pain or tissue damage'.

O'Mahoney proudly says the study has been quoted in several legal cases that arose from coffee-burn incidents, 'sometimes by both sides!'

Lee, Hye-Seong, Earl Carstens, and Michael O'Mahony (2003). 'Drinking Hot Coffee: Why Doesn't It Burn the Mouth?'. *Journal of Sensory Studies* 18 (1): 19–32.
Borchgrevink, Carl P., Alex M. Susskind, and John M. Tarras (1999). 'Consumer Preferred Hot Beverage Temperatures'. *Food Quality and Preference* 10 (2): 117–21.

SMELLS LIKE HOLY SPIRIT

Did early Christians smell inspiration? Susan Ashbrook Harvey's book *Scenting Salvation: Ancient Christianity and the Olfactory Imagination* assures us that yes, they did. The book might, metaphorically, help other academics wake up and smell the stale coffee: here is a pungent research topic that researchers have, until now, hardly bothered to sniff at.

Harvey is a professor of religious studies at Brown University in Providence, Rhode Island. Her 442-page tome explores (1) the most compelling odours, and (2) what and how early Christians thought about those odours.

The table of contents offers topics to tempt even a casual bookstore browser. Read these seven items aloud to a friend or loved one, and you'll see:

- The Olfactory Context: Smelling the Early Christian World
- A Martyr's Scent
- Olfaction and Christian Knowing
- The Smell of Danger: Marking Sensory Contexts
- The Fragrance of Virtue: Reordering Olfactory Experience
- Sanctity and Stench
- Asceticism: Holy Stench, Holy Weapon

At least one other soul is spreading the word about *Scenting Salvation*. Brent Landau, a doctoral student at Harvard Divinity School, published an appreciative essay in the *Bryn Mawr Classical Review*. 'Wake up, fellow scholars!', he virtually shouts. This book 'will be a valuable and provocative resource for scholars working on ancient conceptions of the body, the history of science, ritual studies, asceticism, and Syriac Christianity, among other topics'.

Professor Harvey is not the first to savour early Christian smells, Landau reminds us. But 'her book represents the most comprehensive work on this subject. If there is a watchword that characterizes [Harvey]'s research, it is most certainly "ambiguity"; for not only do odors themselves straddle the line between corporeal and incorporeal, but the judgment of what makes a given smell "good" or "bad" changes drastically for Christian interpreters depending upon the circumstances.'

Modern scholars do not work in hermetically sealed rooms. In addition to writing the book, Harvey gives talks about her work at academic conferences. Among the topics and locations, these have included: 'The Odor of Faith' in Chicago in 1994; 'Sanctity and Stench: When Holy Fragrance Turns Foul' in Michigan in 1994; 'Why the Perfume Mattered: the Sinful Woman in Syriac Exegetical Tradition' at the University of Notre Dame in 1999; later that year, 'On Holy Stench: When the Odor of Sanctity Sickens' in Oxford; and 'Making Sense of Scents: Olfactory Guides for the Late Antique Christian' at Yale in 2006.

She also publishes titbit-packed specialized studies. One of them, 1998's 'St. Ephrem on the Scent of Salvation', tells how, in the fourth century, odours delivered both Good News and Bad News: 'Ephrem emphasizes the experience of smell as the means by which the believer encounters the divine … and learns God's favor or disfavor.'

The study of Christian olfaction once was lost, but now is found. Harvey tells us that, in the fifth century, Christian writers regularly wrote about 'an olfactory dialogue in which human and divine each approached the other through scent'. Professor Harvey has started a dialogue about that dialogue about smells.

Harvey, Susan Ashbrook (1998). 'St. Ephrem on the Scent of Salvation'. *Journal of Theological Studies* 49 (1): 109–28.
Landau, Brent (2007). 'Review: *Scenting Salvation'. Bryn Mawr Classical Review* 2: n.p., http://bmcr.brynmawr.edu/2007/2007-02-44.html.
Harvey, Susan Ashbrook (2006). *Scenting Salvation: Ancient Christianity and the Olfactory Imagination.* Berkeley: University of California Press.

RESEARCH SPOTLIGHT

'CELEBRATING FRAGMENTATION: THE PRESENCE OF ARIS-TOCRATIC BODY PARTS IN MONASTIC HOUSES IN TWELFTH-AND THIRTEENTH-CENTURY ENGLAND'
by Danielle Westerhof (published in *Citeaux Commentarii Cistercienses*, 2005)

A MEASURED LOOK AT CHALK

At long last, after millions of students in thousands of classrooms have freely and incautiously breathed trillions of breaths, there's a report about the question: how much chalk dust enters the air when a teacher uses a blackboard?

The study, 'Assessment of Airborne Fine Particulate Matter and Particle Size Distribution in Settled Chalk Dust during Writing and Dusting Exercises in a Classroom', was done by Deepanjan Majumdar, D.G. Gajghate, Pradeep Pipalatkar and C.V. Chalapati Rao of the National Environmental Engineering Research Institute in Nehru Marg, India.

The team weighed each piece of chalk before and after using it. They collected chalk dust from the air, and also the dust that fell on to a long sheet of paper laid over the base of the blackboard.

Their experiment featured three kinds of chalk, one black-board, an eraser, an aerosol spectrometer (to measure and record the amount of dust floating in the air) and a Cilas model 1180 particle-size analyser.

The researchers tried to ensure maximally pure conditions for the measurements. 'All the windows and the only door were closed airtight', the 'fans present in the classroom were not operated', and 'personal movement in the classroom was completely restricted during the experiment to minimise resuspension of dust from floor'.

The report explains that in schools that still use chalk, teachers brave the greatest direct risk: 'During teaching, entry of chalk dust in the respiratory system through nasopharyngeal region and mouth could be extensive in teachers due to their proximity to the board and frequent opening of mouth during lectures and occasional gasping and heavier breathing due to exhaustion.' As per current state of knowledge on particulate matter vis-à-vis chalk dust, it 'may remain suspended in air for some time before settling on the floor and body parts of the teachers and pupils'.

The scientists acknowledge that chalk and chalkboards these days are being supplanted, in many schools, by whiteboards and other more modern, less intrinsically dusty technology. But chalk still enjoys wide usage in many countries.

The study, published in the journal *Indoor and Built Environment*, ruefully concludes: 'Though real-time airborne chalk dust generation was found to be low in this study ... and did not contain toxic materials, chalk dust could be harmful to allergic persons and may cause lacrimation and breathing troubles in the long run and certainly is a constant nuisance in classrooms as it may soil clothes, body parts, audiovisual aids and study materials.'

Majumdar, Deepanjan, D.G. Gajghate, Pradeep Pipalatkar and C.V. Chalapati Rao (2011). 'Assessment of Airborne Fine Particulate Matter and Particle Size Distribution in Settled Chalk Dust during Writing and Dusting Exercises in a Classroom'. *Indoor and Built Environment* 21 (4): 541–51.

IN BRIEF

'DOES SITTING ON YOUR HANDS MAKE YOU BITE YOUR TONGUE? THE EFFECTS OF GESTURE PROHIBITION ON SPEECH DURING MOTOR DESCRIPTIONS'

by Autumn B. Hostetter, Martha W. Alibali and Sotaro Kita (paper presented at the annual conference of the Cognitive Science Society, 2007)

The authors, who are at the University of Wisconsin–Madison and the University of Birmingham, report: 'Participants in the hands restrained condition were given a 25 x 60 x 2 cm wooden board to place across their laps. On the top of this board, there were several strips of Velcro. The participants were also given cotton gloves to wear that had the opposite side of the Velcro attached to the palms and fingers. They were asked to place their hands on the board, so that the two sides of the Velcro adhered. In this way, they were discouraged from moving their hands during the task without being forcefully restrained ... It seems ... that sitting on your hands does influence your tongue, though it does not make you bite it completely.'

THE PIERCING SOUNDS OF SCIENCE

The sounds of science are in some cases surprising. Or annoying. Or both.

BANG. One of history's most stimulating experiments – stimulating for the volunteer research subjects, at least – was carried out by W. Dixon Ward and Conrad Holmberg of the University of Minnesota. Their summary report, 'Effects of High-Speed Drill Noise and Gunfire on Dentists' Hearing', appeared in a 1969 issue of *the Journal of the American Dental Association*.

The Ward-Holmberg paper is a pleasure to read. It begins: 'A high-pitched, whining noise can be most annoying ...' Concerned that constant exposure to the whirr of a dental drill might be dangerous to the dentist's hearing, Ward and Holmberg decided to investigate.

Their report specifies that 'most of' their testing was done on volunteers. It also mentions that 'two elderly men could not be induced to respond properly to [the] testing procedure.' The scientists also allude to 'a misunderstanding' in eleven cases, wherein data seems to have been gotten from the dentists' wives, rather than from the dentists.

Ward and Holmberg concluded that 'there is only scant evidence that high-speed drills have much influence on the hearing of dentists in Minnesota, especially in comparison to the effects of gunfire.'

BOOM. Margaret Bradley and Peter Lang of the University of Florida meticulously exposed people to what they called 'naturally occurring sounds (e.g., screams, erotica, bombs, etc.)'. These various sounds – there were sixty of them altogether – included beer, cows, a roller coaster, bees, a growling dog, a ticking clock and lovemaking. The volunteers were also shown pictures, including photos of a lamp, an owl, a cow and a burn victim. Drs Bradley and Lang say for the most part their volunteers responded emotionally to sounds pretty much the same way as to pictures. But their report, published in the journal *Psychophysiology*, does note that 'larger startle reflexes were elicited when listening to unpleasant, compared with pleasant, sounds'.

HISS. Some research subjects get a less meaningful, less sharply defined listening experience. In 2002, Tetsuro Saeki, at Yamaguchi University in Ube, Japan, and several colleagues published a report called 'A Method for Predicting Psychological Response to Meaningless Random Noise Based on Fuzzy System Model'. The title speaks for itself. Details, for those who cannot infer them, can be found in the journal *Applied Acoustics*.

YIKES. To the non-specialist, some of this research may sound surprising, annoying or even meaningless. But for someone, somewhere, such reactions may themselves sound like something worth studying.

Ward, W. Dixon, and Conrad J. Holmberg (1969). 'Effects of High-Speed Drill Noise and Gunfire on Dentists' Hearing'. *Journal of the American Dental Association* 79 (6): 1383–7.
Bradley, Margaret M., and Peter J. Lang (2000). 'Affective Reactions to Acoustic Stimuli'. *Psychophysiology* 37 (2): 204–15.
Saeki, Tetsuro, Shizuma Yamaguchi, Yuichi Kato and Kensei Oimatsu (2002). 'A Method for Predicting Psychological Response to Meaningless Random Noise Based on Fuzzy System Model'. *Applied Acoustics* 63 (3): 323–31.

MONKEY, FLOSS

Monkey flossing became a formal practice, at least experimentally, in the late 1970s, thanks to a dentist named Jack Caton. Twenty years later, a physician named David C. Sokal, inspired by the monkey flossing, patented a top/bottom flossing-reminder and floss-dispensing device for humans. Monkeys themselves apparently began unassistedly flossing themselves not long afterwards. But likely those animals did so on their own, not influenced by either Dr Caton's experiment or Dr Sokal's invention.

Caton became the world's foremost monkey flosser in 1979, when he published a small study in the *Journal of Clinical Periodontology*. Based at the Eastman Dental Center in Rochester, New York, he worked with six Rhesus monkeys, all of whom had 'gross amounts of plaque and generalized moderate to severe gingivitis'. ('Gingivitis' is dental lingo meaning 'inflamed gums'.)

Caton tested several methods to improve the monkeys' oral condition. Flossing, brushing and mouthwashing all helped, he reported. No matter what the treatment, the healthiest result came from doing it at least three times a week.

Years later, Sokal saw the Caton recommendations (which,

he points out, 'proved adequate for Rhesus monkeys'), weighed them against other research findings, and concluded that cleaning teeth every second day is 'satisfactory'.

But humans sometimes need reminders. So Sokal invented what he calls a 'floss dispenser with memory aid for flossing upper and lower teeth in separate sessions'.

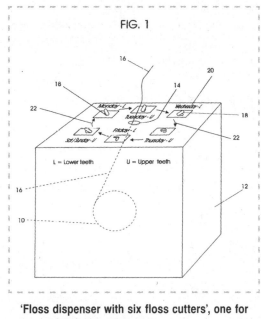

'Floss dispenser with six floss cutters', one for almost every day of the week

Unlike conventional dental floss dispensers, Sokal's has two different clips to slice the floss and hold it in readiness for next time. One is labelled 'Lower teeth – Monday Wednesday Friday', the other 'Upper teeth – Tuesday Thursday Sat/Sunday'. The patent offers variations: electronic day and jaw indicators, and as many as six floss clips (Saturday and Sunday share a single clip).

Monkeys can themselves be inventive. The August 2007 issue of the *American Journal of Primatology* features a report entitled 'Long-tailed Macaques Use Human Hair as Dental Floss'. Written by scientists at Kyoto University, in Japan, and Ubon Rajathanee and Chulalongkorn universities, in Thailand, it builds on a report from the year 2000 that 'two individual *Macaca fascicularis* monkeys in Lopburi, Thailand, used human hair as dental floss'.

The researchers observed similar behaviour with many monkeys, who plucked from sometimes-willing humans. They also learned that, if given human hairpieces, the monkeys in effect used them as floss dispensers, plucking out strands and spinning them into floss. These various monkey achievements are impressive, say the scientists: 'Utilizing women's hair as dental floss is not a simple task; the monkeys need to sort the hair, make a string with it and hold it tightly with both hands to brush their teeth when they feel that pieces of food remain … It was interesting that some monkeys appeared to remove only a few pieces of hair as though they understood that there was an optimum number of hairs required for use as dental floss.'

To see monkeys teach their young how to floss, visit http:// www.dailymotion.com/video/x8n4o3_monkeys-teach-young-to-floss-their_fun#.Udq-RlOxOUk.

Caton, Jack (1979). 'Establishing and Maintaining Clinically Healthy Gingivae in Rhesus Monkeys'. *Journal of Clinical Periodontology* 6 (4): 260–3.
Sokal, David C. (1998). 'Floss Dispenser with Memory Aid for Flossing Upper and Lower Teeth in Separate Sessions and Method'. US patent no. 5,826,594, 27 October 27.
Watanabe, Kunio, Nontakorn Urasopon and Suchinda Malaivijitnond (2007). 'Long-tailed Macaques Use Human Hair as Dental Floss'. *American Journal of Primatology* 69 (8): 940–4.

MAY WE RECOMMEND

'GAGGING DURING IMPRESSION MAKING: TECHNIQUES FOR REDUCTION'

by Sarah Farrier, Iain A. Pretty, Christopher D. Lynch and Liam D. Addy (published in *Dental Update*, 2011)

The authors, at Cardiff University, Wales, report: 'In everyday dental practice one encounters patients who either believe themselves, or subsequently prove themselves, to be gaggers.'

CURED, THERAPEUTICALLY

A 2011 medical study recommends a method called 'nasal packing with strips of cured pork' as an effective way to treat uncontrollable nosebleeds.

Ian Humphreys, Sonal Saraiya, Walter Belenky and James Dworkin at Detroit Medical Center in Michigan treated a girl who had a rare hereditary disorder that brings prolonged bleeding. Publishing in the *Annals of Otology, Rhinology, & Laryngology*, they pack the essential details into two sentences: 'Cured salted pork crafted as a nasal tampon and packed within the nasal vaults successfully stopped nasal hemorrhage promptly, effectively, and without sequelae … To our knowledge, this represents the first description of nasal packing with strips of cured pork for treatment of life-threatening hemorrhage in a patient with Glanzmann thrombasthenia.'

They acknowledge a long tradition of using pork to treat general epistaxis, i.e. nosebleed. The technique fell into disuse, they speculate, because 'packing with salt pork was fraught with bacterial and parasitic complications. As newer synthetic hemostatic agents and surgical techniques evolved, the use of packing with salt pork diminished.'

In 1976, Dr Jan Weisberg of Great Lakes, Illinois, wrote a letter to the journal *Archives of Otolaryngology*, bragging that he, together with a Dr Strother and a Dr Newton, had been 'privileged' to treat a man 'for epistaxis secondary to Rendu-Osler-Weber disease', an inherited problem in which blood vessels develop abnormally. For their patient, the period of hospitalization was ten days and the patient was discharged with salt pork packing still in his nose.

In 1953, Dr Henry Beinfield of Brooklyn, New York, published a treatise called 'General Principles in Treatment of Nasal Hemorrhage'. Beinfield explains: 'Salt pork placed in the nose and allowed to remain there for about five days has been used, but the method is rather old-fashioned.'

In 1940, Dr A.J. Cone of the Washington University School of Medicine in St. Louis, praised the method in a paper called 'Use of Salt Pork in Cases of Hemorrhage'. In Cone's experience, 'it has not been uncommon in the St. Louis Children's Hospital service to have a child request that salt pork be inserted in his nose with the first sign of a nosebleed ... Wedges of salt pork have saved a great deal of time and energy when used in controlling nasal hemorrhage, as seen in cases of leukemia, hemophilia ... hypertension ... measles or typhoid fever and during the third stage of labor'.

Way back in 1927, Dr Lee Hurd of New York published a big thumbs-up in 'Use of Salt Pork to Control Hemorrhage' in the *Archives of Otolaryngology*. Hurd enthused: 'It is hard to say just what the action of the pork is, since several factors are present, namely, pressure, salt, tissue juices and fat... Usually I do not use any outside dressing to hold the pork in place...'

Humphreys, Ian, Sonal Saraiya, Walter Belenky and James Dworkin (2011). 'Nasal Packing with Strips of Cured Pork as Treatment for Uncontrollable Epistaxis in a Patient with Glanzmann Thrombasthenia'. *Annals of Otology, Rhinology, & Laryngology* 120 (11): 732–36.

Weisberg, Jan J. (1976). 'Rendu-Osler-Weber Disease – Is Embolization Beneficial?'. *Archives of Otolaryngology* 102 (6): 385.

Beinfield, Henry H. (1953). 'General Principles in Treatment of Nasal Hemorrhage: Emphasis on Management of Postnasal Hemorrhage'. *Archives of Otolaryngology* 57 (1): 51–9.

Cone, A.J. (1940). 'Use of Salt Pork in Cases of Hemorrhage'. *Archives of Otolaryngology* 32 (5): 941–6.

Hurd, Lee M. (1927). 'Use of Salt Pork to Control Hemorrhage'. *Archives of Otolaryngology* 4 (11): November 1927, p. 447.

AN IMPROBABLE INNOVATION

'INTERNAL NOSTRIL OR NASAL AIRWAY SIZING GAUGE'
by Louise S. MacDonald (US patent no. 7,998,093, granted 2011)

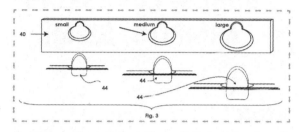

Fig. 3

Gauging small, medium and large digits in place of small, medium and large nostrils

According to the patent, 'certain dimensions of an individual's digits can be correlated to certain dimensions of an individual's internal nostril, or nasal airway size. Gauging the size of an individual's internal nostril or nasal airway by measuring certain dimensions of their digit(s) provides a highly correlating sizing method that is much easier, faster and more convenient than measuring the nostril or nasal airway directly.'

FOUR

BONES, FORESKINS, ARMPITS, SLIME

IN BRIEF

'THE SWEDISH PIMPLE: OR, THOUGHTS ON SPECIALIZATION'
by Jeffrey D. Bernhard (published in the *Journal of the American Academy of Dermatology*, 1995)

Some of what's in this chapter: Explain that itch • Army dandruff • What colour the foreskin • Fingers and pigment, perhaps • Botox for the fetid • Facial hair vs decollatage • Harry Potter, recessively • Kangaroo, harp, infant • Gay anticlockwise scalp hair-whorl rotation • Influence of fingers in eyes • Nude doll with gonorrhoea • Slimeball's gooey power • White hair, too suddenly • Cheek dimples counted and made • Soy sauce hair

THE APPARENT SOURCE OF AN ITCH

'Observations during Itch-Inducing Lecture' is a study published by German researchers in the year 2000. It delivers exactly what the title promises.

Professor Uwe Gieler at Justus-Liebig University in Giessen and two colleagues begin with the basics: 'Itching is defined as a sensation associated with an impulse to scratch.'

They invited people to attend a public lecture called 'Itching – What's Behind It?' The lecture was purportedly recorded for broadcast on television. In fact, the TV cameras were there to record what the audience did – whether people scratched themselves, or failed to scratch themselves.

Method

In cooperation with a public television company which prepared a scientific program on itching, interested persons were invited to participate in an public lecture: 'Itching – what's behind it?'. The aim of the lecturers was to initially present an itch-inducing lecture to the still uninformed audience, followed by a neutral verbal stimulation to feel comfortable. Accordingly, the first part included slides that induce itching

---

Method of soliciting and observing participants for the lecture

The lecture had two parts, the first filled with slides of fleas, mites, scratch marks on skin and other visual stimuli that the scientists hoped would 'induce itching'. The remainder of the lecture presented photos of babies, soft skin and other items meant to 'induce relaxation and a sense of well-being'.

The experiment was to some extent a success. Audience members, on average, scratched themselves more frequently during the itch-provoking part of the lecture. However, because the audience was small, the scientists hesitate to draw any firm conclusion from what happened.

Five years earlier, Clifton W. Mitchell published a treatise called 'Effects of Subliminally Presented Auditory Suggestions of Itching on Scratching Behavior'. It describes his doctoral thesis research at Indiana State University.

For Mitchell, itching and scratching were but a means to an end. His main interest was subliminal perception. His stated intent was 'to create an experimental situation closely analogous to that encountered in commercially available subliminal self-help audiotapes'.

Dr Mitchell had volunteers listen to a specially prepared eleven-minute-long recording of recited 'suggestions of itching'. The report does not specify the nature of these suggestions, other than to say that they were recorded at a level so low that it 'prohibit[ed] conscious detection of spoken words'. Mitchell buried these whisperings beneath a loud soundtrack of what he describes as 'new-age style music'.

A different group of volunteers listened to a recording of just the music.

A third group heard a recording of someone simply, loudly making suggestions about itching. The technical term for this explicit prompting, the report informs us, is 'supraliminal suggestions'.

Mitchell added extra scientific rigour: 'To distract participants from the purpose of the experiment, a biofeedback-type headband with bogus sensors and wiring was fitted around the head.'

He video-recorded the test subjects, then evaluated the recordings to see who scratched themselves, and how often.

The report presents the results of the experiment, condensed into a concise, readable table that is labelled 'Means and standard deviation for number of scratching-type behavior observed for each group across phases'. Those results were a bit paradoxical. Those who listened to the overt itching messages scratched themselves least, those who listened to the pure music scratched most.

There was, Mitchell reports finally, 'no evidence' that listening to subliminally presented auditory suggestions of itching led to an increase in scratching behaviour.

Niemeier, Volker, Jörg Kupfer and Uwe Gieler (2000). 'Observations during Itch-inducing Lecture'. *Dermatology and Psychosomatics* 1 (1 suppl.): 15–8.
Mitchell, Clifton W. (1995). 'Effects of Subliminally Presented Auditory Suggestions of Itching on Scratching Behavior'. *Perceptual and Motor Skills* 80 (1): 87–96.

MAY WE RECOMMEND

'SEVERE INFESTATION OF A SHE-ASS WITH THE CAT FLEA CTENOCEPHALIDES FELIS FELIS (BOUCHE, 1835)'

by I. Yerhuham and O. Koren (published in *Veterinary Parasitology*, 2003)

DANDRUFF IN THE ARMY OF PAKISTAN: A COMPREHENSIVE LOOK

Public knowledge about dandruff in Pakistan's army comes mainly from a study called 'Knowledge, Attitude and Practice regarding Dandruff among Soldiers', written by Naeem Raza, Amer Ejaz and Muhammad Khurram Ahmed, published in 2007 in *the Journal of the College of Physicians and Surgeons – Pakistan.*

Raza, Ejaz and Ahmed surveyed eight hundred male soldiers of all ranks, ascertaining each soldier's knowledge about, and personal experience with, dandruff. The survey was 'designed keeping in mind the general taboos of our region about dandruff, which included visits to doctors, homeopathic physicians or "hakims", use of oils, any home-made remedies or commercial products'.

If this sampling of soldiers was truly representative, we now know that approximately 65 percent of Pakistani soldiers have, or have had, dandruff 'either permanently or periodically'. 'Almost two thirds of the respondents stated to remain tense and embarrassed because of their dandruff.' Noting that the 'media has played an important role in making people think like that', the study concludes with a recommendation. Healthcare professionals should make a greater effort to educate the populace.

KNOWLEDGE, ATTITUDE AND PRACTICE REGARDING DANDRUFF AMONG SOLDIERS

Naeem Raza, Amer Ejaz* and Muhammad Khurram Ahmed**

ABSTRACT

Objective: To assess the knowledge, attitude and practice regarding dandruff among soldiers.

Design: Cross-sectional study.

Place and Duration of Study: Departments of Dermatology at Combined Military Hospitals, Abbottabad, Malir and Gujranwala from January 2006 to March 2006.

Patients and Methods: Serving male soldiers posted at Abbottabad, Malir and Gujranwala Cantonments were included in the study. Convenience sampling was used to distribute the questionnaire. All soldiers included in the study were asked or assisted by the trained staff to fill the close-ended questionnaire in Urdu, which included information about demographic profile and questions relevant to the objectives of the study. The ethical requirements for the study were fulfilled. SPSS-10 was used for data management.

Results: A total of 800 serving male soldiers were surveyed. Five hundred and twenty-one soldiers (65.1%) answered yes to dandruff, whereas 279 (34.9%) replied in negative. Dandruff was considered a disease by 433 (83.1%) respondents. Hair fall (n=392, 75.2%) and scalp itching (n=380, 72.9%) were the common symptoms and 330 (63.3%) respondents were embarrassed by dandruff. Bad water (n=93, 17.8%), winter (n=40, 07.6%) and lack of sleep (n=30, 05.7%) were considered the most common causes of dandruff. Majority of the individuals (n=487, 93.4%) used different hair oils and household remedies for the treatment of dandruff. One hundred and fourteen (21.9%) and 50 (09.6%) participants consulted doctors and traditional healers respectively for their dandruff. Advertisements in electronic or print media and wall hoardings etc. influenced 213 (40.9%) respondents to use various anti-dandruff shampoos, hair tonics and oils.

Conclusion: Dandruff is a common problem and there is a need for education programmes and formulation of a policy regarding the positive role of media on health matters.

KEY WORDS: Dandruff. Pityriasis capitis. Malassezia. Knowledge. Remedies. Soldiers.

Both the numbers and the reactions are typical of the region and the world, according to a study published three years later in the *Indian Journal of Dermatology*, called 'Dandruff: The Most Commercially Exploited Skin Disease'.

The Indian report sketches the underlying situation: 'Dandruff is a common scalp disorder affecting almost half of the population at the pre-pubertal age and of any gender and ethnicity. No population in any geographical region would have passed through freely without being affected by dandruff at some stage in their life.' It helpfully fills in the etymology. 'The word *dandruff* (*dandruff, dandriffe*) is of Anglo-Saxon origin, a combination of "tan" meaning "tetter" and "drof" meaning "dirty" '.

The Pakistan military report cites a 1990 monograph called 'The History of Dandruff and Dandruff in History. A Homage to Raymond Sabouraud'. That homage was written by Didier Saint-Léger of l'Oréal in Aulnay-sous-Bois, France, and published in the journal *Annales de Dermatologie et de Vénéréologie*. Saint-Léger explains that Raymond Jacques Adrien Sabouraud (1864–1938), a French dermatologist, painter and sculptor, is the dominant figure in humanity's effort to understand dandruff.

Saint-Léger shares how dandruff figured in Sabouraud's greatness: 'In one of his books, written at the beginning of this century, Raymond Sabouraud devotes some 280 pages to the history of dandruff. Their reading illustrates how, from the Greeks to Sabouraud's era, this desquamative disease has been subjected to endless doctrinal and scientific conflicts.'

A medical book written during Raymond Sabouraud's lifetime speaks admiringly of the man: 'It is said that Sabouraud can tell your moral character, the amount of your yearly income and what you have eaten for breakfast by looking at the root of one of your hairs.'

Raza, Naeem, Amer Ejaz and Muhammad Khurram Ahmed (2007). 'Knowledge, Attitude and Practice regarding Dandruff among Soldiers'. *Journal of the College of Physicians and Surgeons – Pakistan* 17 (3): 128–31.
Ranganathan, S., and T. Mukhopadhyay 'Dandruff: The Most Commercially Exploited Skin Disease'. *Indian Journal of Dermatology* 55 (2): 130–4.

Saint-Léger, Didier (1990). 'The History of Dandruff and Dandruff in History: An Homage to Raymond Sabouraud'. *Annales de Dermatologie et de Vénéréologie* 117 (1): 23–7.
Thompson, Ralph Leroy (1908). *Glimpses of Medical Europe*. Philadelphia: J.B. Lippincott Company, p. 145.

MAY WE RECOMMEND

'THE HISTORY OF FRECKLES IN ART'
by H.W. Seimens (published in *Der Hautarzt*, 1967)

COLOURFUL FORESKIN RESEARCH

Bryan B. Fuller is the world's top expert on skin colour in human foreskins.

Fuller's foreskin research was for a long time based at the University of Oklahoma. He then became the founder and CEO of DermaMedics (www.dermamedics.com).

A research paper Fuller co-authored with four colleagues in 1990 is the most frequently cited study on the topic of foreskin colour. Entitled 'The Relationship between Tyrosinase Activity and Skin Color in Human Foreskins', it appeared in the *Journal of Investigative Dermatology*. It makes lively reading.

The scientists pre-select their foreskins on the basis of race. The paper explains: 'The race of the child was determined from the race of both parents. Foreskins were only used from children whose parents were *either* racially caucasian or black. No foreskins from racially mixed marriages were used.'

The Fuller process of preparing and utilizing a foreskin is complex.

Seen from the point of view of a foreskin, this is a many-stage adventure. First, the foreskin is surgically removed from its birthplace. Then it is placed on a gauze pad that's been saturated with a fluid called 'Hank's balanced salt solution'. It is then trimmed and sliced into five-square-millimetre chunks. Then each chunk is homogenized three times. It is then sonicated three times. (You may not be familiar with sonication.

Sonication, in the words of the Hielscher Ultrasonics company, which makes sonicators, is 'a very effective method for the mixing, homogenizing, emulsifying, dispersing, disintegrating, and degassing of liquids by means of ultrasonic cavitation'.) The foreskin bits are then frozen, then centrifuged, then sonicated once more.

By this time, the foreskin has been through a lot. But the adventure is really just beginning. Now, at last, the foreskin bits get analysed, but that is a story for another time.

Fuller's patent (US patent no. 5,589,161) for using foreskins to test skin-tanning solutions is the *ne plus ultra* on how to use foreskins to test skin-tanning solutions.

During his academic days, one of Fuller's main aims, according to his website, was 'to develop skin care products which can stimulate melanin production (tanning) in fair-skinned individuals'. Five of his eleven foreskin-related patents, though, were about how to make skin become lighter. One, called 'Method for Causing Skin Lightening', features a 1,300-word exposition about foreskins.

Scientists of an earlier generation fondly recall D.A. Pious and R.N. Hamburger's study of fifty cultures of human foreskin cells, published in 1964. Pious and Hamburger, however, had little to say about the colour of the foreskins.

And of earlier times, there is little on the record. Most disappointing is the fact that foreskin colour is not mentioned at all in Frederick M. Hodges' instant-classic of a report on 'The Ideal Prepuce in Ancient Greece and Rome', which was published in 2001 in the *Bulletin of the History of Medicine*. A Fuller account is wanted.

Fuller, Bryan B. (1996). 'Pigmentation Enhancer and Method'. US patent no. 5,589,161, 31 December.
— (2000). 'Method for Causing Skin Lightening'. US patent no. 6,110,448, 29 August.
Pious, D.A., R.N. Hamburger and S.E. Miles (1964). 'Clonal Growth of Primary Human Cell Cultures'. *Experimental Cell Research* 33 (3): 495–507.
Hodges, Frederick M. (2001). 'The Ideal Prepuce in Ancient Greece and Rome: Male Genital Aesthetics and Their Relation to Lipodermos, Circumcision, Foreskin Restoration, and the Kynodesme'. *Bulletin of the History of Medicine* 75 (3): 375–405.

IN BRIEF

A FULLER PARTIAL ACCOUNTING OF BRYAN B. FULLER'S PATENTS AND PATENTS PENDING:

'Pigmentation Enhancer and Method'.
US patent no. 5,540,914, 30 July 1996.
'Pigmentation Enhancer and Method'.
US patent no. 5,554,359, 10 September 1996.
'Pigmentation Enhancer and Method'.
US patent no. 5,589,161, 31 December 1996.
'Pigmentation Enhancer and Method'.
US patent no. 5,591,423, 7 January 1997.
'Pigmentation Enhancer and Method'.
US patent no. 5,628,987, 13 May 1997.
'Composition for Causing Skin Lightening'.
US patent no. 5,879,665, 9 March 1999.
'Enhancement of Skin Pigmentation by Prostaglandins'.
US patent no. 5,905,091, 18 May 1999.
'Method of Lightening Skin'.
US patent no. 5,919,436, 6 July 1999.
'Method of Lightening Skin'.
US patent no. 5,989,576, 23 November 1999.
'Composition for Causing Skin Lightening'.
US patent no. 6,096,295, 1 August 2000.
'Method for Causing Skin Lightening'.
US patent no. 6,110,448, 29 August 2000.

RESEARCH SPOTLIGHT

'SECOND TO FOURTH DIGIT RATIO, SEXUAL SELECTION, AND SKIN COLOUR'

by John T. Manning, Peter E. Bundred and Frances M. Mather (published in *Evolution and Human Behavior,* 2004)

The authors, at the University of Central Lancashire and the University of Liverpool, write: 'Skin pigment may be related to mate choice, marriage systems, resistance to micoorganisms, and photoprotection. Here we use the second to fourth digit ratio (2D:4D) to disentangle the relationships.'

IN EXPENSIVELY GOOD ODOUR

Botox – a/k/a 'botulinum toxin' – has had a curious reputation with the public. First it was feared: it can kill, after all. Then it was cheered: fashionable *femmes et hommes* were delighted to hear that something with a hint of danger could make their wrinkles vanish. Now we are on the verge of a third and rather different wave of acclaim.

For a long time, only horror film fans, physicians and hypochondriacs were lovingly familiar with the basics about botulinum toxin. Everyone else would hear mention of it only occasionally – whenever the food-borne illness botulism struck down some unhappy soul. The US Centers For Disease Control and Prevention put out a concise description of the illness and its cause: 'Botulism is a rare but serious paralytic illness caused by a nerve toxin that is produced by the bacterium *Clostridium botulinum.*'

As all up-to-date celebrity worshippers know, one particular variety of botulinum toxin – called 'botulinum toxin A' – turned out to be useful in a cosmetically valuable way. Botulinum toxin A has other medical uses, too. One of them, the control of excessive, by-the-bucketful underarm sweating, inspired the notion that botulinum toxin might be useful in combating nasty armpit odour.

The notion was put to the test using T-shirts, sniff tests and volunteers who allowed doctors to inject botulinum toxin A into one armpit and a salt solution into the other.

This is specialized research, and discussing it calls for a bit of specialized vocabulary. Bromidrosis is a word familiar to physicians, to pedants and to some of the people who suffer from bromidrosis. It is especially familiar to those sufferers who have consulted a pedant or a physician. Bromidrosis means 'fetid or foul-smelling perspiration'. The word axillary means 'having to do with the armpit'.

I mention these two obscure words – bromidrosis and

axillary – because Drs Marc Heckmann, Bianca Teichmann and Bettina M. Pause of the Ludwig-Maximilian-University in Munich, and Gerd Plewig of Christian-Albrecht-University in Kiel, use them in describing their volunteers. The report says that 'although the volunteers had no history of bromidrosis, the axillary odor was clearly rated as unpleasant prior to treatment'.

T-shirt sniff tests were performed before, and again seven days after, the botox-in-the-armpit injections. The results, report the doctors, were dramatic: 'Apart from reduced odor intensity, axillae treated with botulinum toxin A were also rated as smelling less unpleasant or literally more pleasant, which means an improvement in the quality of body odor. Presently, any explanation for this phenomenon can only be highly speculative.'

We see here the birth of a tentative new rule of thumb: what doesn't kill you makes you smoother, and less stinky.

Heckmann, Marc, Bianca Teichmann, Bettina M. Pause, and Gerd Plewig (2003). 'Amelioration of Body Odor After Intracutaneous Axillary Injection of Botulinum Toxin A'. *Archives of Dermatology* 139 (1): 57–9.

IN BRIEF

'SELF-REFERENT PHENOTYPE MATCHING IN A BROOD PARASITE: THE ARMPIT EFFECT IN BROWN-HEADED COWBIRDS (MOLOTHRUS ALTER)'
by Mark E. Hauber, Paul W. Sherman and Dóra Paprika
(published in *Animal Cognition,* 2000)

LOSING BY A WHISKER

Not being a barber, and not having had an adulthood that spanned 130 years, Dwight E. Robinson was in no position to report firsthand the frequency of changes in relative prevalence of sideburns, moustaches and beards in London during the years 1842–1972. He used an indirect source: issues of the *Illustrated London News* published during that time.

Robinson, a business professor at the University of Washington in Seattle, gathered his findings about those findings into a study that he called 'Fashions in Shaving and Trimming of the Beard: The Men of the *Illustrated London News*, 1842–1972'. It was published not in Britain but in the *American Journal of Sociology*, in 1976.

Robinson used and, he says, improved upon a general analytic technique pioneered by Jane Richardson and Alfred L. Kroeber, who in 1940 'measured annual fluctuations in width and length of skirts, waistlines and decolletage as ratios to women's heights'.

'My procedure for gathering data', Robinson explains, 'was, quite literally, to take a head count, determining for any one year the comparative frequencies of men's choices among five major features of barbering: sideburns alone, sideburns and moustache in combination, beard (a category that included any amount of whiskers centring on the chin), moustache alone, and clean shavenness'. For the mathematically inclined, Robinson notes that 'the number of clean-shaven men in any year is by definition the reciprocal of the sum of those in the four whisker categories'.

Intent on choosing data that would accurately reflect the reality of Londoners' facial hair, Robinson excluded photos of groups (because some faces might appear only partially, or in misleading angles), royalty (because royals receive more press coverage, if not necessarily more hair, than the general populace), advertisements and 'pictures of non-Europeans'. One graph shows, beginning in the year 1885, a stark, almost unceasing rise in clean-shavenness.

Sideburns decline until about the year 1920, thereafter making only negligible appearances. Beards, too, hit bottom in 1920, but quasi-periodically grow back to modest popularity.

In that hair-oilshed year 1921, moustaches reach their all-time peak, adorning nearly 60 percent of the non-grouped, non-royal, non-advertised, non-non-European men appearing in the *Illustrated London News*. Thereafter, moustaches dominate all other forms of facial hair.

In one provocative graph, Robinson plots two grand, 115-year-long, rising-and-falling waves. One represents women's skirt width (in proportion to the women's height). The other shows the pervasiveness of beards among the male population. The skirt-width-ratio wave precedes the beard wave by a gap of twenty-one years. Robinson says the data reveals that 'men are just as subject to fashion's influence as women'.

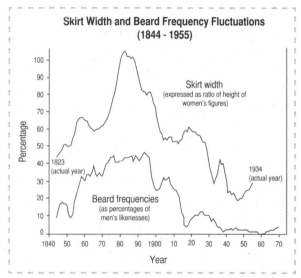

The monograph notes: 'Skirt width (1823–1934) and beard frequency fluctuations (1844-1955), five-year moving averages. The time scales of the two curves have been positioned to allow for assumed 21-year lead in skirt fluctuations possibly related to comparative youthfulness of subjects'.

Fashion tells just part of the story. A quarter century after Robinson's analysis, an independent, aptly-named scholar – the Irish-born, America-adopted Nigel Barber – published a study in which he reports that: 'Men shave their mustaches, possibly to convey an impression of trustworthiness, when the marriage market is weak'.

Robinson, Dwight E. (1976). 'Fashions in Shaving and Trimming of the Beard: The Men of the *Illustrated London News*, 1842–1972'. *American Journal of Sociology* 81 (5): 1133–41.

Richardson, Jane, and Alfred L. Kroeber (1940). 'Three Centuries of Women's Dress Fashions: A Quantitative Analysis.' *Anthropological Records* 5 (2): 111–54.

Barber, Nigel (2001). 'Mustache Fashion Covaries with a Good Marriage Market for Women'. *Journal of Nonverbal Behavior* 25 (4): 261–72.

MAY WE RECOMMEND

'HARRY POTTER AND THE RECESSIVE ALLELE'
by J.M. Craig, R. Dow and M. Aitken (published in *Nature*, 2005)

'DUTY OF CARE TO THE UNDIAGNOSED PATIENT: ETHICAL IMPERATIVE, OR JUST A LOAD OF HOGWARTS?'
by Erle C.H. Lim, Amy M.L. Quek and Raymond C.S. Seet (published in the *Canadian Medical Association Journal*, 2006)

KANGAROO CARE

You might think that an Israeli Medical Association report called 'Combining Kangaroo Care and Live Harp Music Therapy in the Neonatal Intensive Care Unit Setting' is the first medical study of the combined effects, on newborns, of kangaroo care and music therapy. Not so.

The invention of kangaroo care (also called kangaroo therapy) is widely attributed to a pair of doctors in Colombia in the late 1970s. Initially, both the idea and the name triggered scepticism. Thus the appearance in 1990 of a paper called 'Kangaroo Care: Not a Useless Therapy', in a magazine published by America's National Association of Neonatal Nurses.

The idea of kangaroo care is for premature babies to spend most of their time being held or pressed against the mother, the two maintaining direct 'skin-to-skin' contact. This was meant as a substitute for incubators in places where those were unavailable. Later, some doctors and nurses began recommending that even the most modern hospitals adopt the practice.

Eventually came attempts to see whether – or not – kangaroo care produces good effects. Research journals began publishing studies, including the provocative 'Kangaroo Care Modifies Preterm Infant Heart Rate Variability in Response to Heel Stick Pain: Pilot Study'.

Then someone got the idea of adding music. Researchers at several institutions in Taiwan combined forces to perform an experiment, documented in a 2006 paper, 'Randomized Controlled Trial of Music during Kangaroo Care on Maternal State Anxiety and Preterm Infants' Responses', in the *International Journal of Nursing Studies*. The experimenters had mothers and premature babies snuggle skin-to-skin while listening to recorded music emanated, for sixty minutes, from a Philips AZ-1103 'ghetto blaster'. At the same time, other mothers and preemies neither snuggled skin-to-skin nor heard recorded music.

The Taiwan researchers found that the kangaroo'ed babies slept a bit more than the non-kangaroo'ed babies, and cried a bit less. And, they say, the kangarooing mothers gradually felt ever-so-slightly lessened anxiety. As for the babies' health – the main reason people recommend kangaroo care – they reported 'no significant difference on infants' physiologic responses' between those who got kangarooing and music, and those who did not.

The new Israeli kangaroo-plus-harp-music study also reports that their particular 'combined therapy had no apparent effect on the tested infants' physiological responses or behavioural state'. But a similar study – which the Israeli study does not mention – done in Finland and published online a year earlier by the *Nordic Journal of Music Therapy* reported that kangaroo therapy accompanied by live harp music 'did' affect the medical state of the child. The Finns say that it 'decreased the pulse, slowed down the respiration and increased the transcutaneous O_2 saturation', and 'affected the blood pressure significantly'.

And so doctors and nurses must await further research before they know the value of prescribing kangaroo care with live harp music.

Schlez, Ayelet, Ita Litmanovitz, Sofia Bauer, Tzipora Dolfin, Rivka Regev and Shmuel Arnon (2011). 'Combining Kangaroo Care and Live Harp Music Therapy in the Neonatal Intensive Care Unit Setting'. *Israeli Medical Association Journal* 13 (6): 354–8.
Jorgensen, K.M. (1999). 'Kangaroo Care: Not a Useless Therapy'. *Central Lines* 15 (3): 22.
Cong, Xiaomei, Susan M. Ludington-Hoe, Gail McCain and Pingfu Fu (2009). 'Kangaroo Care Modifies Preterm Infant Heart Rate Variability in Response to Heel Stick Pain: Pilot Study'. *Early Human Development* 85 (9): 561–7.
Lai, Hui-Ling, Chia-Jung Chen, Tai-Chu Peng, Fwu-Mei Chang, Mei-Lin Hsieh, Hsiao-Yen Huang and Shu-Chuan Chang (2006). 'Randomized Controlled Trial of Music During Kangaroo Care on Maternal State Anxiety and Preterm Infants' Responses'. *International Journal of Nursing Studies* 43 (2): 139–46.
Teckenberg-Jansson, Pia, Minna Huotilainen, Tarja Pölkki, Jari Lipsanen and Anna-Liisa Järvenpää (2011). 'Rapid Effects of Neonatal Music Therapy Combined with Kangaroo Care on Prematurely-Born Infants'. *Nordic Journal of Music Therapy* 20 (1): 22–42.

SWIRLING SEXUAL POSITIONS

Amid the swirls of controversy that buffet other sexuality researchers, one man focuses, quietly, on swirls. In a report called 'Excess of Counterclockwise Scalp Hair-Whorl Rotation in Homosexual Men', Dr Amar J.S. Klar announces a subtle discovery. 'This is the first study', he writes, 'that shows a highly significant association of biologically specified counterclockwise hair-whorl rotation and homosexuality in a considerable proportion of men in samples enriched in gays.' Klar heads the developmental genetics section of the Gene Regulation and Chromosome Biology Laboratory at the US National Cancer Institute in Frederick, Maryland. His hair-swirl study appears in a 2004 issue of the *Journal of Genetics*.

The phenomenon is easy to overlook. Klar explains: 'Since the hair whorl is found at the top ("crown") of the head and thereby it is difficult to observe one's own whorl and the direction of orientation is seemingly an unimportant feature, most people are oblivious to the direction of their hair-whorl rotation. It takes two mirrors to observe one's own hair-whorl.'

His monograph includes a photograph showing the 'scalp hair whorl of an anonymous man selected from the general public', and directs the reader to hold that picture in front of a mirror in order to 'appreciate the counterclockwise orientation'.

How difficult is it to collect hair-whorl-direction data? Klar says that he, for one, got lucky: 'By chance I happened to be vacationing at a beach where preponderance of gay men was fortuitously noticed. The subjects were considered to be homosexuals because of their public display of stereotypical interpersonal relationship deemed typical of homosexual men. This assessment was reinforced by the dearth of females and children on the beach … Conveniently, the gay men were highly concentrated in one area of the beach. Such considerations made it relatively easy to collect the data on groups of predominantly gay men with great confidence even though the subjects were not asked for their sexual preference.'

A year later Klar returned to the same beach and collected another load of data. He reports: 'Altogether in a combined sample of 272 mostly gay men observed, 29.8% exhibited counterclockwise hair-whorl orientation'. This, he says, is 'vastly different from the value of 8.4% counterclockwise rotation found in the public at large, which included both males and females.'

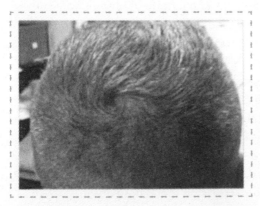

Observed: 'hair-whorl phenotype'. The author explains: 'By holding the picture in front of a mirror and looking at its image, the reader can appreciate the counter-clockwise orientation.'

The study does not take account of the erstwhile hair-whorl directionality of persons who are now bald. He explicitly excluded them from consideration, along with anyone who was wearing a sun hat.

Klar suggests a direction for further exploration: 'It should be equally interesting to compare the proportions of clockwise and counterclockwise hair-whorl orientations in lesbian women with those in females at large.'

The report ends with a simple notice that deftly fends off the research-is-a-waste-of-government-money crowd: 'Author's personal funds were used for the study.'

Klar, Amar J. S. (2004). 'Excess of Counterclockwise Scalp Hair-Whorl Rotation in Homosexual Men'. *Journal of Genetics* 83 (3): 251–5.

RESEARCH SPOTLIGHT

'DIGIT RATIO (2D:4D) IN LITHUANIA ONCE AND NOW: TEST-ING FOR SEX DIFFERENCES, RELATIONS WITH EYE AND HAIR COLOR, AND A POSSIBLE SECULAR CHANGE'
by Martin A. Voracek, Albinas Bagdonas, and Stefan G. Dressler (published in *Collegium Antropologicum*, 2007)

EXPERIMENTS WITH NUDE DOLLS

A generic life-size doll, with no modifications, was the key element in at least one unplanned experiment – the experiment documented in a 1993 letter to the journal *Genitourinary Medicine* entitled 'Transmission of Gonorrhoea through an Inflatable Doll'. But, generally, scientists who conduct planned experiments that rely on life-size dolls prefer to carefully optimize, or even create, their own doll.

That unplanned inflatable doll experiment centred on a ship's captain who 'with some hesitation … told the story' while being treated at a sexual disease clinic in Greenland. The captain had without permission entered an absent crewman's cabin, borrowed a piece of equipment and later suffered the consequences.

That inflatable doll was not purpose-built for scientific use. Only through delightful happenstance did it satisfy the scientists', as well as the captain's, needs. Most scientists hate to depend on serendipity, especially if they have to depend on a doll.

A study called 'Convective Heat Transfer from a Nude Body Under Calm Conditions: Assessment of the Effects of Walking With a Thermal Manikin' exhibits the forethought and niggling care that can go into acquiring a suitable nude doll. Five mechanical engineers at the University of Coimbra, Portugal, wanted to study how, as a person strolls in the open air, heat flows both away from and into the skin. So, they obtained 'a Pernille type thermal mannequin named Maria', which 'is articulated and divided into 16 parts independently controlled by a computer'. Maria features 'a fibreglass armed polyester shell covered with a thin nickel wire wound around all the body to ensure heating and temperature measurement'.

Jintu Fan · Xiaoming Qian
New functions and applications of Walter, the sweating fabric manikin

In 2004, an entire, special issue of the *European Journal of Applied Physiology* featured twenty-seven studies involving mannequins. In some of those studies, the researchers refer to their mannequin by name. Jintu Fan and Xiaoming Qian, of Hong Kong Polytechnic University's institute of textiles and clothing, called their monograph 'New Functions and Applications of Walter, the Sweating Fabric Manikin'.

Fan and Qian write that Walter 'simulates perspiration using a waterproof, but moisture-permeable, fabric "skin" [that] can be unzipped and interchanged with different versions to simulate different rates of perspiration'. Fan and Qian say their greatest challenge about Walter 'is to measure the amount of water added to or lost from [him].'

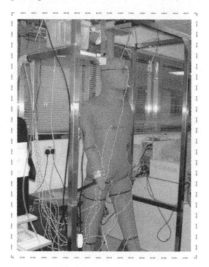

'Walter in "walking motion" '

In other experimental studies, ones where a mannequin is subjected to hellacious treatment, the writing sometimes shows a particular, uncomfortable kind of restraint. 'Exposure to hot water steam is a potential risk in the French Navy', says one such paper, explaining a moment later that 'this extreme environment during an accident leads to death in a short time'. In that study, as in others involving extreme exposures, the mannequin's name – if anyone bothered even to give it a name – is withheld from the public.

Ellen Kleist and Harald Moi were honoured with the 1996 Ig Nobel Prize in public health for the doll ghonorrhoea report.

Kleist, Ellen, and Harald Moi (1993). 'Transmission of Gonorrhoea through an Inflatable Doll'. *Genitourinary Medicine* 69 (4): 322.

Virgílio, A., M. Oliveira, Adélio R. Gaspar, Sara C. Francisco and Divo A. Quintela (2012). 'Convective Heat Transfer From a Nude Body Under Calm Conditions: Assessment of the Effects of Walking With a Thermal Manikin'. *International Journal of Biometeorology* 56 (2): 319–32.

Fan, Jintu, and Xiaoming Qian (2004). 'New Functions and Applications of Walter, the Sweating Fabric Manikin'. *European Journal of Applied Physiology* 92 (6): 641–4.

Desruelle, Anne-Virginie, and Bruno Schmid, 'The Steam Laboratory of the Institut de Médecine Navale du Service de Santé des Armées: A Set of Tools in the Service of the French Navy'. *European Journal of Applied Physiology* 92 (6): 630–5.

AN IMPROBABLE INNOVATION

'HUMAN-FIGURE DISPLAY SYSTEM'

a/k/a mannequin with repositionable arms and legs, by Rebecca J. Bublitz and Annette L. Terhorst (US patent no. 6,601,326, granted 2003)

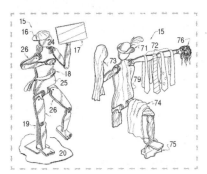

Figures 1 and 8 from US patent no. 6,601,326.

SLIME TO THE RESCUE

Slime would become the US military's prime weapon to immobilize large ships under a scheme outlined for the US Air Force's Air Command and Staff College.

Lieutenant Commander Daniel Whitehurst, a student at the college, figured out how to combine a raft of existing technologies to produce the officially 'non-lethal' armament he calls The Slimeball. He prepared a report in 2009. 'The Slimeball', Whitehurst writes, 'is a two-part weapon system consisting of a floating sticky foam barrier that will resist attempts to remove it, and a submerged gel barrier that will impede movement through a ship channel. The parts can also be used independently of each other.'

Whitehurst gives three examples of targets well suited for The Slimeball's gooey power. He explains how to use it against pirates in Boossaaso, Somalia; against the Iranian navy near the city of Bandar Abbas in the Strait of Hormuz; and against China's underground submarine base at Sanya, on Hainan Island.

The Slimeball requires foam with particular qualities. Whitehurst specifies: 'The primary component of such a material would contain properties commonly found in shaving cream … As commercially formulated, shaving cream is too insubstantial to create more than a nuisance to vessels, but in a denser form and combined with the chemical properties like those of a pre-existing substance known in defence circles as "sticky foam", it would pose a far greater challenge for removal and have a greater dissuasive effect on vessels operating on the surface.'

Sticky foam, Whitehurst explains, was designed for use against people. He allows that it has 'some significant drawbacks' as an anti-personnel weapon. These include 'the risk of suffocation and the inability to transport the target due to the, well, stickiness of the material'.

Whitehurst expresses optimism that this kind of officially non-lethal tool need not be lethal. 'It has been suggested that due to the maturity of knowledge and development in this field, the drawbacks can be "engineered out" ', he writes.

Detail from 'Report Documentation Page'

The US Air Force has a history of imaginative weaponry proposals. Whitehurst says 'many attempts have been made over the years to impede naval forces in a variety of manners, including floating smoke pots, entanglement devices, and even "floating purple mountains of shaving cream" … but none ever made it into wide use.'

More flashily, plans prepared long ago at the US Air Force's Wright laboratory, in Dayton, Ohio, called for development of a potent chemical weapon. This is the so-called gay bomb that makes enemy soldiers sexually irresistible to each other. Wright laboratory was awarded the Ig Nobel peace prize in 2007 for instigating that line of research.

Whitehurst, Daniel L. (2009). 'The Slimeball: The Development of Broad-scale Maritime Non-lethal Weaponry'. Research report submitted in partial fulfillment of graduation requirements, Air Command and Staff College, Air University, Maxwell Air Force Base, Alabama, April.
n.a. (1994). 'Harassing, Annoying, and "Bad Guy" Identifying Chemicals'. Wright Laboratory, WL/FIVR, Wright Patterson Air Force Base, Dayton, Ohio, 1 June.

AND HIS HAIR TURNED WHITE ...

In a study called 'Sudden Whitening of the Hair: An Historical Fiction?', Anne-Marie Skellett, George Millington and Nick Levell, at Norfolk and Norwich University Hospital, try to chop off a myth at its roots. People's hair, they believe, does not all of a sudden turn white. It just doesn't. Goodbye, ye hoary tales of Queen Marie Antoinette of France and Sir Thomas More of England each turning whitehaired the night before being beheaded.

Hair whitening – 'canities' in medical lingo – takes longer than days or even weeks, they report in a 2008 issue of the *Journal of the Royal Society of Medicine*. If somebody's hair did suddenly turn white, they say, it would most likely have an unnatural cause: 'the washing out, or lack of access to a temporary hair dye'.

They suggest one other possible, though maybe nonexistent, mechanism. The disease alopecia totalis makes people's hair fall out. Perhaps, in someone of mixed white and dark hair, some rare form of the disease might make only the dark strands fall out.

In medical monographs over the past one hundred years, doctors have almost uniformly expressed scepticism. In 1972, Josef Jelinek, of New York University medical school, debunked dozens of supposedly documented sudden-hair-whitening claims with a monograph in the *Bulletin of the New York Academy of Medicine*.

Jelinek, an MD, began by directing the blame: 'It was not the physician but rather the historian who first seized on stories of sudden whitening to dramatize the tribulations of famous persons, principally in their grief or fear, in order to interest and astonish the reader. The poet, too, found poignancy in this phenomenon.' He mentions as an example Shakespeare's *King Henry IV*, Part 1, where Falstaff says to Hotspur: 'Worcester is stolen away tonight. Thy father's beard is turned white with the news.'

SUDDEN WHITENING OF THE HAIR*

J. E. JELINEK, M.D.

Clinical Associate Professor of Dermatology
New York University Medical School
New York, N. Y.

THE phenomenon of sudden blanching of the hair has been the subject of debate and speculation for hundreds of years. Vehement assertions both of its occurrence and impossibility are found in the medical literature of the past three centuries. However, it was not the physician but rather the historian who first seized on stories of sudden whitening to dramatize the tribulations of famous persons, principally in their grief or fear, in order to interest and astonish the reader. The poet, too, found poignancy in this phenomenon. Thus Sir Walter Scott in Marmion:

> For deadly fear can time outgo,
> And blanch at once the hair.

Dr Alexander Navarini and Dr Ralph Trüeb at University Hospital of Zürich, Switzerland, have become the foremost modern scholars of sudden-whitening. In 2009, they published a report called 'Marie Antoinette Syndrome', about a fifty-four-year-old woman whose 'entire scalp hair suddenly turned white within a few weeks'. The next year, they published a paper called 'Thomas More Syndrome', about a fifty-six-year-old man 'with sudden total whitening of scalp hair and eyebrows within weeks'.

They potter on about the names: 'Saint Thomas More, who turned white in 1535, ought to have the right of seniority over the Queen of France who succumbed to the same fate in 1793. Since there seems to be no other particular reason for favouring Marie Antoinette over Thomas More, out of fairness, it would seem appropriate to use the term "Marie Antoinette syndrome" for the condition afflicting women and "Thomas More syndrome" for men.' They further contributed to our knowledge of saints' hair with a 2010 paper called 'Beneath the Nimbus'.

Summary of Saints, Their Peculiarities, and the Symbolism of Their Hair

	Mary Magdalene	Perpetua	Onuphrius	Mary of Egypt	Thomas More
Period	First century	Died 203	320-400	344-421	1478-1535
Sanctity	Follower of Christ and the Apostles	Martyr	Hermit	Hermit	Martyr
Account	New Testament (Luke)	Tertullian of Carthage	Paphnutius	Sophronius	History of England
Attributes	Long, un-covered hair, ointment vase	In arena, usually together with Felicity	Wild man completely covered with hair, loin girdle of leaves	Long hair covering naked body	Book, axe
Patronage	Hair stylists	Martyrs	Weavers	Penitents	Statesmen and politicians
Hair condition	Long, beauti-ful hair	Hair in martyrdom	Generalized hypertri-chosis	Long, disheveled hair	Sudden whitening of hair
Symbolism	Female attractiveness	Dignity	Withdrawal from worldly concerns and vanities	Withdrawal from worldly concerns and vanities	Extreme psycho-logical stress

Some saints summarized by hirsuteness, from 'Beneath the Nimbus'

Navarini and Trüeb also disseminated some dark, happy hair news. In a monograph called 'Reversal Of Canities' they explain that they can't really explain why a sixty-seven-year-old 'otherwise healthy' man 'presented with spontaneous repigmentation of his grey hair'.

Skellett, Anne-Marie, George W.M. Millington and Nick J. Levell (2008). 'Sudden Whitening of the Hair: An Historical Fiction?'. *Journal of the Royal Society of Medicine* 101(12): 574–6.

n.a. (1910). 'Sudden Whitening of the Hair'. *The Lancet* 175 (4525): 1430–1.

n.a. (1973). 'Sudden Whitening of the Hair'. *British Medical Journal* 1 (5852): 504.

Jelinek, Joseph E. (1972). 'Sudden Whitening of the Hair'. *Bulletin of the New York Academy of Medicine* 48 (8): 1003–13.

Navarini, Alexander A., Stephan Nobbe and Ralph M. Trüeb (2009). 'Marie Antoinette Syndrome'. *Archives of Dermatology* 145 (6): 656.

Trüeb, Ralph M., and Alexander A. Navarini (2010). 'Thomas More Syndrome'. *Dermatology* 220: 55–6.

Navarini, Alexander A., and Ralph M. Trüeb (2010). 'Why Henry III of Navarre's Hair Probably did Not Turn White Overnight'. *International Journal of Trichology* 2 (1): 2–4.

— (2010). 'Reversal of Canities'. *Archives of Dermatology* 146 (1): 103–4.

Trüeb, Ralph M., and Alexander A. Navarini (2010). 'Beneath the Nimbus: The Hair of the Saints'. *Archives of Dermatology* 146 (7): 764.

IN BRIEF

'WHY THE LONG FACE?: THE MECHANICS OF MANDIBULAR SYMPHYSIS PROPORTIONS IN CROCODILES'

by Christopher W. Walmsley, Peter D. Smits, Michelle R. Quayle, Matthew R. McCurry, Heather S. Richards, Christopher C. Oldfield, Stephen Wroe, Phillip D. Clausen and Colin R. McHenry (published in *PLoS ONE,* 2013)

'Bite, shake and twist' pressure points of three crocodile species

GREEK CHEEK

How many Greek children have dimpled cheeks? Until recently no one really knew, but now there is detailed information as to exactly how many do, how many don't, and where the dimples are.

Athena Pentzos-DaPonte of Aristotle University in Thessaloniki and an international team counted the dimples on 14,141 male and 14,141 female Greek children and adolescents. To be thorough, they observed the children smiling and also not smiling.

The scientists performed this count in 1980. A quarter century later, they finished their quantitative analysis of the data, and published a report in the *International Journal of Anthropology*. The report does not explain the significance, if any, of the number 14,141.

The data paint a clear picture. Approximately 13 percent of Greek children had a noticeable cheek dimple or dimples. Girls and boys were almost equally well dimpled.

Pentzos-DaPonte and her colleagues – Alessandro Vienna from the University of Rome, Larry Brant from the National Institute on Aging in Baltimore, Maryland, and Gertrude Hauser from the University of Vienna – also considered location. Was a dimple on the left? Was it on the right? Or (to put it in technical terms) was the dimpling bilateral? Left-dimpled Greek children were as common as right-dimpled, but only about 3.5 percent of youngsters had dimples in both cheeks.

In 1983 Pentzos-DaPonte and colleague Silke Grefen-Peter published a study of cheek dimpling in Greek adults. About 34 percent of the adults had dimples – almost triple the occurrence of dimpling in Greek youths. The scientists now speculate that most of those adult dimples are literally old-fashioned: the skin aged, lost elasticity and gained sag.

Greece has always been a nation that prizes knowledge. Few other countries, though, know the prevalence of cheek dimpling among any age group. It would seem a straightforward procedure to count dimples. But, increasingly, such figures are becoming subject to manipulation.

For example, Dr Pichet Rodchareon, of the Pichet Plastic Surgery Clinic in Bangkok, Thailand, advertises his willingness to insert cheek dimples for a cost (as I am writing this) of US $1500 per dimple. Rodchareon is one of the most prominent advertisers of this particular service (and other services too; his website says the practice is 'The Leading Aesthetic Plastic Surgery Center of Cosmetic Surgery and Sex Change Surgery in Bangkok, Thailand'), but many plastic surgeons have the skill and experience to sculpt a dimple. Some even offer a reversible cheek dimple operation.

At the 2002 World Congress of Cosmetic Surgery, held in Shanghai, China, Dr Xuan Cuong Nguyen of Vietnam presented a talk called 'An Easy and Precise Way to Make a Cheek Dimple'. This is the same Dr Nguyen who just the year before was awarded the 'World Leader of Cosmetic Surgery' gold cup at a ceremony in Mumbai, India. So far, his method has not proved as easy as it sounds, given that the price of artificial cheek dimples has not yet tumbled. That's unhappy news for the dimple seeker on a tight budget.

Pentzos-DaPonte, Athena, Alessandro Vienna, Larry Brant and Gertrude Hanser (2004). 'Cheek Dimples in Greek Children and Adolescents'. *International Journal of Anthropology* 19 (4): 289–95.
Pentzos-Daponte, Athena (1986). '4 Anthroposcopic Markers in the Northern Greece Population: Hand Folding, Arm Folding, Tongue Rolling and Tongue Folding'. *Anthropologischer Anzeiger* 44 (1): 55–60.

MAY WE RECOMMEND

'THE TREATMENT DILEMMA CAUSED BY LUMPS IN SURFERS' CHINS'

by Jun Fujimura, Kenji Sasaki, Tsukasa Isago, Yasukimi Suzuki, Nobuo Isono, Masaki Takeuchi and Motohiro Nozaki (published in *Annals of Plastic Surgery*, 2007)

HAIR-RAISING TREATS

Alexander Tse-Yan Lee – or, as he generally identifies himself: Alexander Tse-Yan Lee, BH Sci., Dip. Prof. Counsel., MAIPC, MACA – was in the news some years ago, albeit tangentially. For a while the Internet was slightly a-twitter (though not via Twitter, which did not yet exist) with mentions of Lee's piece of writing entitled 'Hair Soy Sauce: A Revolting Alternative to the Conventional'. It appeared in the *Internet Journal of Toxicology*, adding to that publication's stock of grim, occasionally grimy delights. The article's section headings do a good job of both stoking and satisfying the reader's interest:

- The Soy Sauce – An Introduction
- The Cheap Soy Sauce That Aroused the Public
- The Stunning Alternative to Soy – the Human Hair
- Toxic Consequences of The Hair and The Chemicals
- The Boycott Phenomena
- Conclusion

Little attention has been paid to Dr Lee himself, or to his other work.

Lee's stated affiliation is unusual: Queers Network Research of Hong Kong, China. So are his other published papers, several of which also appear in the *Internet Journal of Toxicology*.

'The Foods From Hell: Food Colouring', is filled with colourfully tasty details. It, too, reveals its gist to the reader who skims the section titles:

- Food Colouring Agents: Synthetic Versus Natural
- Coloured Chinese Steamed Corn-Buns
- Coloured Dry Shrimps (Dried Shrimps?)
- Coloured Fruits
- Coloured Vegetables
- Coloured Dark Rice
- Coloured Traditional Chinese Herbal Medicinal Products
- Conclusion

Lee's biography mentions a study called 'It Is Foods that Look Good Kill You', which is described as being 'in press' in the *Internet Journal of Toxicology*. It is not clear, if one goes looking for the report itself, whether 'It Is Foods that Look Good Kill You' was an early title that was later published, to little acclaim, as 'The Foods From Hell', or whether it has yet to appear.

Lee did publish 'Faked Eggs: The World's Most Unbelievable Invention'. It, too, features tiny, tale-telling section titles that, in another setting, might comprise an entire short poem:

- A Brief Introduction to Problem Foods in Mainland China
- The Eggs that Cause Problems
- The 'Red Yolk' Eggs
- The Soil-Filled Eggs
- The Human-Made Eggs
- Is it a good advice to sniff the eggs only?
- Conclusion

Earlier, Dr Lee wrote a book called *My Weight Loss Diary eBook*, which foreshadows many of his later works. He offered it on the Internet. The summary alone may be worth the $6 price: 'With certificate from a pancake challenge after finishing a stack of pancake in less than 2 minutes; holding a record of eating 8 family size pizzas in a buffet dinner; having two burgers with milk shake for lunch everyday and finishing an extra large size frozen chicken with chips and gravy for snacks while still demanding for more.'

Some of the articles disappeared from the Internet, having lived quiet lives that persist in the memories of people who voraciously consume all news about hair, soy sauce or faked eggs – and persist also, of course, in the mind of Alexander Tse-Yan Lee, BH Sci., Dip. Prof. Counsel., MAIPC, MACA.

Lee, Alexander Tse-Yan (2005). 'Hair Soy Sauce: A Revolting Alternative to the Conventional'. *Internet Journal of Toxicology* 2 (1): n.p.

— (2005). 'The Foods From Hell: Food Colouring'. *Internet Journal of Toxicology* 2 (2): n.p.

— (in press). 'It Is Foods that Look Good Kill You'. *Internet Journal of Toxicology.*

— (2005). 'Faked Eggs: The World's Most Unbelievable Invention'. *Internet Journal of Toxicology* 2 (1): n.p.

FIVE

THINKING, ON YOUR FEET AND FINGERS

IN BRIEF

'ARM WRESTLING FRACTURES: A HUMERUS TWIST'
by J.H. Whitaker (published in *American Journal of Sports Medicine*, 1977)

Some of what's in this chapter: Pedestrians in space • Topsy-turvey sloth • Man's heavy leg • Tables and chairs on the highway • Wok twists • Mosh motion • Sock not inside • Manipulation by armadillo • Giving the finger new meaning • Missing this, missing that • Fake foot indications • The knifing of pumpkins • Approaches to cutting flesh • Wealth via cockroach leg • Foot tippling

PEDESTRIAN RESEARCH

As you walk city streets, frustrated at why those other pedestrians behave so frustratingly, be aware that people are trying to improve the situation, but are making progress only in slow steps.

Dr Taku Fujiyama, one of the modern masters in this endeavour, is a lecturer at University College London's Secret Centre, or 'the £17m international centre for PhD training in security and crime-related research'.

Fujiyama began his work before joining Secret. In 2005, while affiliated (as he continued to be) with UCL's Centre for Transport Studies, he wrote a study called 'Investigating Use of Space of Pedestrians'. It proposed a series of experiments in and on a mobile lighting-and-sound-system-equipped 'elevated demountable paved platform'.

Different kinds of people – old, young, fat, thin – ambled and strode along the platform. A laser tracking system monitored their every motion and stoppage.

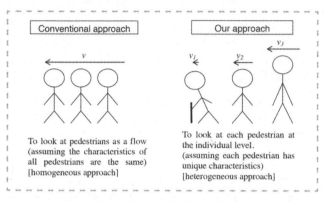

Comparison of two pedestrian approaches

One early experiment focused on how people avoid collisions. Fujiyama observed that 'collision avoidance of a pedestrian reflects his/her spatial requirements for the walking space'.

In a later paper, 'Free Walking Speeds on Stairs: Effects of Stair Gradients and Obesity of Pedestrians', Fujiyama and his colleague Nick Tyler report that they 'did not find any significant difference between the walking speeds of normal and overweight (or moderately obese) participants'.

Fujiyama and Tyler are following the decades-old footsteps of UCL's Ivor B. Stilitz. In the late 1960s, Stilitz analysed rush-hour crowds moving and not moving through ticket halls in five London Underground stations. He published a summary filled with detailed tables and diagrams, including a squiggles and numbers depiction of the 'clouds' that form near a set of ticket machines.

Stilitz sprinkled his paper with a drop of dry engineering humour. He wrote: 'Not surprisingly, the length of queue was correlated with the number of people in it.'

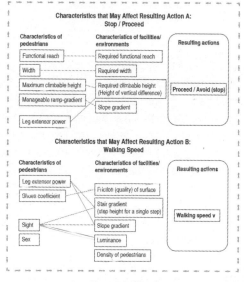

A pedestrian flowchart

Fujiyama, Taku (2005). 'Investigating Use of Space of Pedestrians', Accessibility Research Group, Centre for Transport Studies, University College London, working paper, January.

— (2011). 'Free Walking Speeds on Stairs: Effects of Stair Gradients and Obesity of Pedestrians'. In Peacock, Richard D., Erica D. Kuligowski and Jason D. Averill (eds). *Pedestrian and Evacuation Dynamics*. New York: Springer, pp. 95–106.

Kitazawa, Kay, and Taku Fujiyama (2010). 'Pedestrian Vision and Collision Avoidance Behaviour: Investigation of the Information Process Space of Pedestrians Using an Eye Tracker'. In Klingsch, Wolfram W.F., Christian Rogsch and Andreas Schadschneider (eds). *Pedestrian and Evacuation Dynamics 2008*. New York: Springer, pp. 95–108.

Stilitz, Ivor B. (1969). 'The Role of Static Pedestrian Groups in Crowded Spaces.' *Ergonomics* 12 (6): 821–39.

MAY WE RECOMMEND

'DOLL SHOES: THE CAUSE OF BEHAVIOURAL PROBLEMS?'

by A.K. Leung (published in the *Canadian Medical Association Journal*, 1984)

Doll shoes: the cause of behavioural problems?

I would like to report my experience with a child who had quite a marked behavioural change recently following exposure to a pair of offensive-smelling plastic shoes from a Cabbage Patch doll (Coleco Industries Inc. Hartford Connecticut)

The author, at Alberta Children's Hospital in Calgary, reports: 'One evening, while relatives were visiting from Toronto, his mother slept with him. After 20 minutes in bed she started to feel very nauseated, felt as though she were floating and had a severe headache. At the same time she could smell a very strong odour in the bed, and when she rolled over she realized that the shoes on the doll were the cause of the odour and of her feeling ill. The shoes were discoloured and the stench was very strong. The woman put the shoes away and since that time has noticed a marked improvement in her child's behaviour.'

SLOTHFUL

The name of the sloth is synonymous with a certain style of sin. But scientists pursue them for other reasons, too. The animals move – something they do on occasion – in what can seem mysterious ways. They hang upside down from tree limbs, and sometimes amble that way there. On the ground, ambling right-side-up is their preferred way to get from here to slightly over there. They often snooze.

In 2010 a study called 'Three-dimensional Kinematic Analysis of the Pectoral Girdle during Upside-down Locomotion of Two-toed Sloths' appeared in the journal *Frontiers in Zoology*. John Nyakatura and Martin Fischer of Friedrich-Schiller-Universität in Jena, Germany, analysed the 'suspensory quadrupedal locomotion' of two sloths. They concluded that an earlier biologist had exaggerated, but only slightly, in proclaiming that 'of all mammals, the sloths have probably the strangest mode of progression'.

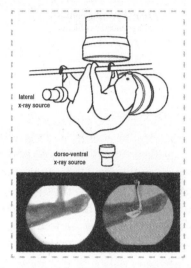

Sloths, recorded, by Nyakatura and Fischer (2010)

Others hesitated less to say more. A 2007 book called *Victorian Animal Dreams: Representations of Animals in Nineteenth-Century Literature and Culture* gets a bit slothful about facts. It says: 'As the early Victorians well knew, the living descendants of the sloth family not only spend their lives suspended upside down in trees, but are also incapable of walking, and are, in fact, so slow that moss grows on their fur.'

There has been more careful research. In 1981, Frank Mendel of the State University of New York, Buffalo, published a painstaking analysis called 'Use of Hands and Feet of Two-toed Sloths (*Choloepus hoffmanni*) during Climbing and Terrestrial Locomotion' in the *Journal of Mammalogy*. Four years later, Mendel published a not-quite-identical paper called 'Use of Hands and Feet of Three-toed Sloths during Climbing and Terrestrial Locomotion'.

A German/Swiss/Panamanian/American team studied slothly sleep. They published a paper in 2008 called 'Sleeping Outside the Box: Electroencephalographic Measures of Sleep in Sloths Inhabiting a Rainforest'.

Sloths do, now and then, get around. Robert Enders published a paper in 1940 called 'Observations on Sloths in Captivity at Higher Altitudes in the Tropics and in Pennsylvania'. Enders transported two sloths from their native Panama to his workplace at Swarthmore College in Pennsylvania, where they soon died.

Sloths can shock experts. A 1989 paper called 'Agonistic Behaviour by Three-toed Sloths' gives this eyewitness account: 'A three-toed sloth (*Bradypus variegatus*) rapidly ascended a cecropia tree, fought briefly and intensely with another adult male, and descended even more rapidly … The social system, visual abilities, and activity budgets of sloths are probably more complex than previously thought.'

In 2010, University of Helsinki researchers examined sloths' hair. They found that a variety of green algae grows there, and that other scientists 'have observed a wide range of animals [there], e.g. moths, beetles, cockroaches and roundworms'.

A German/Peruvian team of scientists published a study in 2011 called 'Disgusting Appetite: Two-toed Sloths Feeding in Human Latrines'.

Evidence: 'A sloth inside the latrine'

Then there's always the big question. In 1971, a study titled 'Why Are Sloths So Slothful?' investigated whether sloths' nerves and brain are sloppily slow. Its conclusion: 'We cannot verify our initial suspicion.'

Nyakatura, John A., and Martin S. Fischer (2010). 'Three-dimensional Kinematic Analysis of the Pectoral Girdle during Upside-down Locomotion of Two-toed sloths (*Choloepus didactylus*, Linné 1758).

Rauch, Alan (2007). 'The Sins of Sloths: The Moral Status of Fossil Megatheria in Victorian Culture'. In Morse, Deborah Denenholz, and Martin A. Danahay (eds). *Victorian Animal Dreams: Representations of Animals in Victorian Literature and Culture*. Aldershot, Hampshire: Ashgate Publishing.

Mendel, Frank C. (1981). 'Use of Hands and Feet of Two-toed Sloths (*Choloepus hoffmanni*) during Climbing and Terrestrial Locomotion'. *Journal of Mammology* 62 (2): 413–21.

— (1985). 'Use of Hands and Feet of Three-toed Sloths (*Bradypus variegatus*) during Climbing and Terrestrial Locomotion'. *Journal of Mammology* 66 (2): 359–66.

Rattenborg, Niels C., Bryson Voirin, Alexei L. Vyssotski, Roland W. Kays, Kamiel Spoelstra, Franz Kuemeth, Wolfgang Heidrich and Martin Wikelski (2008). 'Sleeping Outside the Box: Electroencephalographic Measures of Sleep in Sloths Inhabiting a Rainforest'. *Biology Letters* 4 (4): 402–5.

Enders, Robert K. (1940). 'Observations on Sloths in Captivity at Higher Altitudes in the Tropics and in Pennsylvania', *Journal of Mammology* 21 (1): 5–7.

Green, Harry (1989). 'Agonistic Behavior of Three-toed Sloths (*Bradypus variegatus*)'. *Biotropica* 21 (4): 369–72.

Suutari, Milla, Markus Majaneva, David P. Fewer, Bryson Voirin, Annette Aiello, Thomas Friedl, Adriano G. Chiarello and Jaanika Blomster (2010). 'Molecular Evidence for a Diverse Green Algal Community Growing in the Hair of Sloths and a Specific Association with *Trichophilus welckeri* (Chlorophyta, Ulvophyceae)'. *BMC Evolutionary Biology* 10: 86.

Heymann, Eckhard W., Camilo Flores Amasifuén, Ney Shahuano Tello, Emérita R. Tirado Herrera and Mojca Stojan-Dolar (2011). 'Disgusting Appetite: Two-toed Sloths Feeding in Human Latrines', *Mammalian Biology* 76 (1): 84–6.

Toole, James F. (1971). 'Why Are Sloths So Slothful?', *Transactions of the American Clinical and Climatological Association* 82: 131–5.

RESEARCH SPOTLIGHT

'THE WEIGHT OF THE LEG IN LIVING MEN'
by Robert Bennett Bean (published in the *American Journal of Physical Anthropology*, 1919)

CHAIR RULES

The phrase 'Tables and chairs on the highway' has a uniformly accepted meaning in all of England and Wales.

That meaning is legalistic, deriving, we are told, from part VIIA, section 115 (A to K) of the Highways Act 1980, a chunk of parliamentary prose that has the title 'Provision of Amenities on Certain Highways'. In describing those amenities, though, it makes no mention – none at all – of chairs or tables or any other kind of common furniture. The phrase 'Tables and chairs on the highway' appears nowhere – nowhere – in Highways Act 1980.

Nevertheless, many regional and local authorities proclaim that part VIIA, section 115 (A to K) of the Highways Act 1980 – devoid though it is of tables and chairs – gives them authority to regulate all aspects of civic life that are covered by the phrase 'Tables and chairs on the highway'. And regulate it they do:

> *Chelmsford borough council* publishes a document called 'Guidelines for Placing Tables and Chairs on the Highway under Section 115 Part VIIA of the Highways Act 1980'.
>
> *Westminster city council* goes with the shorter title, 'Guidelines for the Placing of Tables and Chairs on the Highway'.
>
> *Eastleigh borough council* keeps it terse; they call theirs 'Tables and Chairs on the Highway'.

What do all the tables and chairs on the highway regulations regulate? Cafes, restaurants, pubs, bars and shops that wish to place tables and chairs outside, on the street.

The regulations mainly deal with safety, trying to ensure 'that free and safe passage for pedestrians can be maintained'. For some councils – among them Basingstoke, Bath and North-east Somerset, Oxfordshire, and Kensington and Chelsea – that's about the extent of it.

Other councils have larger concerns. They care, deeply, about the furniture. In the royal borough of Windsor and Maidenhead, the council lets it be known that the 'type and colours' of all tables and chairs 'need approval'.

Similarly, Rushmoor borough council, in Hampshire, says: 'Upholstered chairs, cushions and similar effects will not normally be considered acceptable ... Materials, patterns, colours and style of furniture ... must not be too bright, loud or garish (Reason: public and visual amenity) ... Variation in design, eg chairs with or without arms, will only be acceptable if from the same design range and of the same general style (Reason: public and visual amenity).'

The almost-neighbouring Eastleigh borough council says: 'Be a good neighbour! ... Only furniture approved by the council may be used ... Full details, including metric dimensions, materials and colours, of proposed furniture, ideally accompanied by photographs, illustrations or drawings, will be required as part of the application ... The colour of furniture should be attractive but not too bright, garish or overly reflective.' (Eastleigh borough council also says: 'The crockery and cutlery used in street cafes should be of good quality and a uniform style.')

Up in Lambeth, however, the council simply says: 'The type and style of furniture is your choice.'

These policies are all available, at the time I'm writing this, on the official websites of the various councils. Should one of them disappear from the Net, you would likely be welcome to visit the council office and invite the helpful person there to join you in a local cafe – one that has placed chairs and tables on a highway – for a cup of tea and a civilized chat.

STANDARDIZED WOK MEN

'The Effect of Wok Size and Handle Angle on the Maximum Acceptable Weights of Wok Flipping by Male Cooks', a report in the journal *Industrial Health*, does more than its title reveals. It also shows how to standardize an intricate physical test.

Many professional wok-users use a big one. Almost all of those woks have a straight handle. That's bad, say Swei-Pi Wu and Cheng-Pin Ho at Huafan University, and Chin-Li Yen at National Pingtung University of Science and Technology, Taiwan. '[We found that] a small wok [about thirty-six centimetres across] with an ergonomically bent handle is the optimal design, for male cooks, for the purposes of flipping.'

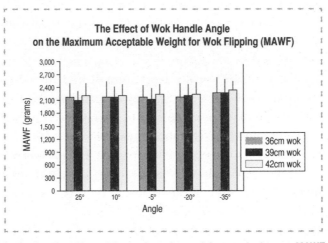

Reviewing the effect of five wok angles and three wok sizes on MAWF

Professional wok-weilding cooks are prone to shoulder, neck, lower back/waist and finger/wrist aches and injuries. Wok-flipping brings some glaring risk: 'The repeated action of swinging the wok up and down, to quickly stir the food in the wok, involves extensive arm and wrist movement, especially dorsi flexion, palmar flexion and wrist radial and ulnar deviation. This non-neutral posture, accompanied by high torque and a high rate of repetition is very apt to cause cumulative trauma disorder injuries in the user's upper extremity.'

Wu, Ho and Yen had twelve experienced Chinese cooks repeatedly flip woks of three different sizes, with handles at five different angles. The tricky part was standardizing the repetitions for all those people over all those wok-size-and-handle-angle combinations.

Their main tool used was a loudspeaker. They required the cooks to follow a strict protocol. The central part ran like this:

Flip the wok nine times, adding or removing soybeans if the wok feels too light or too heavy.

The loudspeaker says: 'Adjust the weight, for the last time!' Flip once more.

The loudspeaker says: 'Please hold the culinary spatula!'

The loudspeaker says 'Ready!', then two seconds later says: 'Begin!'

Lift the wok and shake it, three times, with the non-dominant hand, lifting the culinary spatula with the dominant hand, 'to perform the simulated food stir-frying task, first from right to left and then from left to right and from front to back, three times in total'. Then put the wok down. Repeat this cycle eight times.

The loudspeaker says: 'Please empty soybeans into the container on the left!'

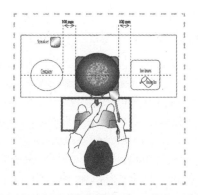

The 'experimental layout', as viewed from above

Wu is an old hand at creating such tests. In 1995 he and a colleague published a study called 'Effects of the Handle Diameter and Tip Angle of Chopsticks on the Food-serving Performance of Male Subjects'. He and a different colleague had earlier published a treatise called 'An Investigation for Determining the Optimum Length Of Chopsticks'.

Those two studies left some gaps in our knowledge of chopstick optimization. A researcher named Tam Chan filled those gaps in 1999 with a paper called 'A Study for Determining the Optimum Diameter of Chopsticks'.

Wu, Swei-Pi, Cheng-Pin Ho and Chin-Li Yen (2011). 'The Effect of Wok Size and Handle Angle on the Maximum Acceptable Weights of Wok Flipping by Male Cooks'. *Industrial Health* 49 (October): 755–64.
Wu, Swei-Pi (1995). 'Effects of the Handle Diameter and Tip Angle of Chopsticks on the Food-serving Performance of Male Subjects'. *Applied Ergonomics* 26 (6): 379–85.
— Chan, Tam (1999). 'An Investigation for Determining the Optimum Length of Chopsticks'. *International Journal of Industrial Ergonomics* 23 (1/2): 101–5.

MAY WE RECOMMEND

'COLLECTIVE MOTION OF HUMANS IN MOSH AND CIRCLE PITS AT HEAVY METAL CONCERTS'

by Jesse L. Silverberg, Matthew Bierbaum, James P. Sethna and Itai Cohen (published in *Physical Review Letters*, 2013)

COLD ADVICE: PUT A SOCK ON IT

Socks over shoes surpass shoes over socks for strolling on slippery city slopes, says a study done in New Zealand. In other words – in the words of the study itself – 'wearing socks over shoes appears to be an effective and inexpensive method to reduce the likelihood of slipping on icy footpaths'.

Lianne Parkin, Sheila Williams and Patricia Priest conducted an experiment to test the wisdom of a local winter tradition. The trio, based at the University of Otago in Dunedin, published a report in the *New Zealand Medical Journal*. They explain: 'There are anecdotal reports that pedestrians who wear socks over top of their footwear are less likely to slip and fall in icy conditions. Advocates of this practice include our local council (in Dunedin) which advises residents who prefer to walk (rather than drive) in icy conditions to "put a pair of old socks over your shoes to increase grip" '.

Their city has some famously hilly sections that grow treacherous come wintertime: 'Damp weather followed by freezing conditions can transform a quick journey to work into a lengthy and perilous expedition.' They 'initially considered recruiting volunteers to walk down a short suburban street (Baldwin Street) which, according to the *Guinness Book of Records*, is the steepest street in the world'. But legal and other considerations led them instead to send people down two other streets, with merely San Francisco-grade inclinations.

Parkin, Williams and Priest found it simple to recruit volunteers: 'To be eligible for inclusion in the trial, passing pedestrians simply needed to be travelling in a downhill direction. It was decided *a priori* that persons already wearing socks over their shoes would not be eligible.'

The research team documented every fall, and wrote comments (such as 'walked confidently', 'clung to fences or parked cars', 'crawled') about the demeanour of each volunteer during their descent.

Not all experiments give clear results. This one did. 'Wearing socks over footwear significantly reduced the self-reported slipperiness of icy footpaths and a higher proportion of sock-wearers displayed confidence in descending the study slopes. The only falls occurred in people who were not wearing (external) socks.'

Sock inside, on ice

But despite the safety advantage, wearing one's socks over one's shoes can create or exacerbate a problem. The problem is of a social nature.

In 1989, two researchers extracted gossip from a group of young (aged seven to eleven) American schoolchildren, asking each child to discuss the reputations of each of their classmates. The kids prattled on about behaviours that, to them, were warning signs of weirdness: 'eats like a pig, bangs head on desk, sounds like a car, fidgety, acts like a monster, wears socks over shoes'.

The what-other-people-will-think problem cropped up in the Dunedin shoes-over-socks study. Parkin, Williams and Priest note that: 'although participants in the intervention group were told that they could keep their socks, many (who appeared to have image issues) opted to return them to the outcome assessors – including one young man who promptly fell on leaving the assessment area.'

Parkin, Williams and Priest were awarded the 2010 Ig Nobel Prize in physics for their work.

Parkin, Lianne, Sheila M. Williams, and Patricia Priest (2009). 'Preventing Winter Falls: A Randomised Controlled Trial of a Novel Intervention'. *New Zealand Medical Journal* 122 (1298): 31–8.
Rogosch, Fred A., and Andrew F. Newcomb (1989). 'Children's Perceptions of Peer Reputations and Their Social Reputations Among Peers'. *Child Development* 60 (3): 597–610.

MAY WE RECOMMEND

'A BAYESIAN APPROACH TO WIGGLE-MATCHING'
by J.A. Christen and C.D. Litton (published in the *Journal of Archaeological Science*, 1995)

SHIFTY DAMAGE DONE

Archaeologists know that the ground they examine can be literally rather shifty. The reasons for this shiftiness can be disturbing, beastly and even childish.

For many an archaeologist, the greatest treasures are arti-facts – 'objects produced or shaped by human workmanship', as the dictionary puts it. The exact location of an artifact can be as important as the thing itself. In a dirt heap, what is next to what – and especially, what is on top of or below what – can impart telling information.

Archaeologists have it tough, though, because things do move around underground. Sometimes this is due to the actions of burrowing animals. Over the years, archaeologists have cautioned each other to beware of ants, termites, earthworms, and, even more famously, rats and other rodents.

A 2003 report from South America points out that: 'There is, however, one animal that despite its regional ubiquity and notable burrowing behavior has received little attention in the archaeological literature: the armadillo.'

1. Are armadillos responsible for substantial vertical displacement of archaeo-logical materials? If so, is there preferential direction of movement (e.g., more artifacts going up than down)?
2. Can they mix different layers of the same deposit? If so, what is the magnitude and nature of mixing?
3. Do the displaced artifacts show any spatial pattern, either vertically or hori-zontally?
4. Do the displaced artifacts present any modal value or threshold for weight, size, or shape?
5. Is it possible to recognize the action of armadillos from the way the displaced pieces are arranged?

Detail: Araujo and Marcelino's research questions

Writing in the *Geoarchaeology*, Astolfo Gomes de Mello Araujo, of the Universidade de São Paulo, and his colleague José Carlos Marcelino go into considerable detail. Living in an age of specialization, they focus on one aspect of the arma-dillo question: 'Although it seems clear that armadillos can move archaeological materials upward', they write, 'previous studies have not considered whether they also move artifacts downward.'

The scientists conducted an experiment with armadillos in

an enclosed underground dirt heap at the São Paulo zoo. They stacked coloured, flaked stones and bits of ceramic into layers, then let the animals have their way.

The armadillos proved to be impressive object-movers. Araujo and Marcelino proudly report: 'Some of our findings have never been reported in the literature, such as the fact that armadillos can translocate artifacts downward to great depths as well as expel them towards the surface.'

To a naïve non-archaeologist, such concerns may seem like child's play. But for good archaeologists, child's play is sometimes a serious problem.

Norman Hammond and the then infant Gawain Hammond illustrated this by doing an experiment. The older Hammond is a professor, now based at Boston University. The Hammonds' report, 'Child's Play: A Distorting Factor in Archaeological Distribution', appeared in 1981.

The elder Hammond created an artificial trash pile one metre in diameter. He stocked it with wine jars, liquor bottles and twelve mostly-empty beer cans. The younger Hammond was then permitted to play briefly in the pile. The report concludes: 'The interpolation of "child-play" may profoundly modify the initial archaeological pattern, and transmit it into an arbitrary pattern with an unrelated structure. The causes of this change must be allowed for in investigations of artifact.'

For measuring how the course of history, or at least the contents of an archaeological dig site, can be scrambled by the actions of a live armadillo, Araujo and Marcelino were awarded the 2008 Ig Nobel Prize in archaeology.

Araujo, Astolfo Gomes de Mello, and José Carlos Marcelino (2003). 'The Role of Armadillos in the Movement of Archaeological Materials: An Experimental Approach'. *Geoarchaeology* 18 (4): 433–60.
Hammond, Gawain, and Norman Hammond (1981). 'Child's Play: A Distorting Factor in Archaeological Distribution'. *American Antiquity* 46 (3): 634–6.

IN BRIEF

'THE PROBLEM OF THE LOCOMOTIVE-GOD'
by W.S. Taylor and E. Culler (published in the *Journal of Abnormal and Social Psychology*, 1929)

SUCCESS AT YOUR FINGER TIPS

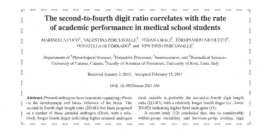

The second-to-fourth digit ratio correlates with the rate of academic performance in medical school students

MARINELLA COCO[1], VALENTINA PERCIAVALLE[2], TIZIANA MACI[3], FERDINANDO NICOLETTI[1], DONATELLA DI CORRADO[2] and VINCENZO PERCIAVALLE[1]

Departments of [1]Physiological Sciences, [2]Formative Processes, [3]Neurosciences, and [4]Biomedical Sciences, University of Catania, Catania; [5]Faculty of Sciences of Formation, University of Kore, Enna, Italy

Received January 3, 2011; Accepted February 15, 2011

DOI: 10.3892/mmr.2011.456

Abstract. Prenatal androgens have important organizing-effects on the development and future behavior of the brain. The second-to-fourth digit length ratio (2D:4D) has been proposed as a marker of these prenatal androgen effects, with a relatively longer fourth finger indicating higher prenatal androgen ... most suitable is probably the second-to-fourth digit length ratio (2D:4D), with a relatively longer fourth finger (i.e., lower 2D:4D) indicating higher fetal androgens (11).

A recent study (12) concluded that, due to considerable within-group variability and between-group overlap, digit ...

A 2011 Italian study points towards a handy way to identify who might become a top doctor. The method is simple: compare the lengths of a person's second finger and fourth finger. The technique is disclosed in the monograph 'The Second-To-Fourth Digit Ratio Correlates with the Rate of Academic Performance in Medical School Students' and appears in the May-June issue of the journal *Molecular Medicine Reports*.

Researchers who spend time studying this finger ratio – the individuals who have found meaning by looking to both sides when someone shows them a middle finger – call it '2D:4D'. That's shorthand for the phrase 'second-to-fourth digit length ratio'. The big dog, the Einstein of the field, is Professor John Manning of the University of Liverpool, whose work has included how to spot footballing stars of the future by looking at finger length (for more on Manning and other high achievers in the field, see page 131).

The authors of the Italian study, mostly medical researchers from the University of Catania in Sicily, and Kore University in Enna, write: '2D:4D has been shown to predict the success of men who play sports and of financial traders.' But, they claim, their paper is the first to reveal what 2D:4D says about high flyers in a highly competitive university system such as the state-run Italian medical schools.

After measuring the fingers of forty-eight male medical students using callipers accurate to 0.2 millimetres, they concluded that students with a slightly lower 2D:4D ratio of the right hand (although, confusingly, it is not quite clear whether this means the fingers are more similar or different in length) are more likely to be successful. The researchers noted in particular how those who passed their medical school admissions test had a significantly lower 2D:4D ratio of the right than those who failed. Interestingly, however, finger ratios could not predict how those students performed in their exams.

It's possible that identifying future stars by examining their fingers is even better than choosing people at random (see page 293). But until and unless somebody does the research, no one can say for sure.

Coco, Marinella, Valentina Perciavalle, Tiziana Maci, Ferdinando Nicoletti, Donatella Di Corrado and Vincenzo Perciavalle (2011). 'The Second-to-Fourth Ratio Correlates with the Rate of Academic Performance in Medical School Students'. *Molecular Medicine Reports* 4: 471–6.
Manning, John T., and Rogan P. Taylor (2001). 'Second to Fourth Digit Ratio and Male Ability in Sport: Implications for Sexual Selection in Humans'. *Evolution Human Behavior* 22 (1): 61–69.

IN BRIEF

'POINTING THE WAY: THE DISTRIBUTION AND EVOLUTION OF SOME CHARACTERS OF THE FINGER MUSCLES OF FROGS'
by Thomas C. Burton (published in *American Museum Novitates*, 1998)

THE MEANING OF THE FINGER

Most people who have done research on fingers have either measured them or tried to repair them.

Among them was V. Rae Phelps of the University of Texas

and Tulane University in New Orleans. In 1952, Phelps compiled a capsule history of early finger findings, waxing nearly poetic about researchers' many missteps and mistakes. In his influenced monograph 'Relative Index Finger Length as a Sex-influenced Trait in Man', published in the *American Journal of Genetics,* he reports: 'Ecker (1875) noted that three manifestations of relative finger length may be discerned in the living model: index finger shorter than ring finger; index finger equal in length to ring finger; and index finger longer than ring finger. Many of the earlier workers failed to recognize this variability'.

He went on to name names. 'Gerdy (1829) stated that the index finger is always shorter than the ring finger, while according to Carus (1853) and Humphry (1861), the index finger exceeds the ring finger in length', he notes. 'Langer (1865) declared that the index finger is shorter than or nearly equal to the ring finger. Alix (1867), Grining (1886), Baker (1888), Schultz (1926), and Wood-Jones (1920, 1941) point out that although the index finger is usually shorter than the ring finger, it may in certain circumstances exceed the length of the ring finger'.

Phelps concentrated mostly on the lengths of fingers and the failures of his predecessors, choosing not to speculate on what, if anything, fingers themselves might mean.

Thankfully, we are experiencing a golden age in finger investigations. Some pioneers began to wonder, 'What do fingers mean?' Modern finger researchers see profound implications in the relative-finger-length possibilities that Ecker (and Phelps) pointed out, most especially, 2D:4D ratios. Here is the big idea, in a nutshell:

1) The body's many hormones are involved in many things during foetus-hood, childhood and adolescence. These many things happen at various times and in various ways.

2) Each hormone involved in these things has many different effects. Scientists have noticed some of

these effects, and understand a few of them, at least a little bit.

3) Testosterone is one of the hormones that scientists have noticed.

4) Testosterone may somehow, at some time, affect how long fingers grow.

5) The relative lengths of a person's fingers may say something about how much testosterone was in the body at some point earlier in their life.

6) The amount of testosterone in a person's body at some point earlier in their life affects lots of other things.

This big idea is often credited to Professor John T. Manning. Thus his status as the large canine, the genius of the field. But fingers are hot in the research world, and Manning is not alone in eyeing them.

Emma Nelson, a researcher and visiting lecturer at the University of Chester who sometimes works with Professor Manning, has her own series of finger-related studies. One examines the 2D:4D ratio of hand outlines stencilled on cave walls. Nelson hopes that 2D:4D will show whether the hands belonged to ancient cave men or to cave women. Another study uses finger-length ratio to better understand the topic 'testes size and dominance in a Group of Captive Chimpanzees'.

Martin Voracek of the University of Vienna (featured on page 299) takes a manifold interest in fingers. Sometimes, he collaborates with Professor Manning, but he also does research independently.

Drs Voracek and Manning recently reported that, by measuring a man's 2D:4D, they could predict, to some degree, how many sexual partners the man would have. But, they say, this works for heterosexual men, not for homosexuals. So far, they express confidence only that it applies to Austrian men.

Voracek and Manning also published a report on the perhaps inevitable question of whether the finger length ratio is related to the length of the penis. Yes, they say, it is.

Teaming with other colleagues, Voracek also published a study about how 2D:4D relates to the performance of skilled fencers. (Manning turned his own attention to 2D:4D and surfers.)

S. Marc Breedlove, a professor of neuroscience at Michigan State University, has become a giant of finger research. In 2000, his study called 'Finger-Length Ratios and Sexual Orientation' appeared in the journal *Nature*. It features a handy graph. The reader can see that, on average, gay and straight women have distinctly different finger ratios – and that gay and straight men each have their own distinctly differing ratios. These distinctions, curiously, are true of the right hand or of the left, but not of both.

Breedlove made a more specific discovery two years later. He and several colleagues published a study called 'Differences in Finger Length Ratios Between Self-identified "Butch" and "Femme" Lesbians'. Their findings are summed up in this sentence: 'We surveyed individuals from a gay pride street fair and found that lesbians who identified themselves as "butch" had a significantly smaller 2D:4D than did those who identified themselves as "femme".'

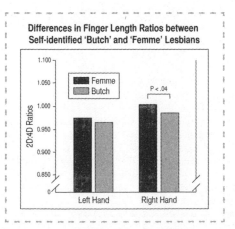

One proposed meaning of the finger

There are also finger financialists. Professor Siegfried Dewitte is an applied economist at Katholieke Universiteit Leuven. Bram van den Bergh trained under Dewitte. Their work, compared to some of their competitors, is less abstruse. For instance, in 'Second to Fourth Digit Ratio and Cooperative Behavior', published in *Biological Psychology* in 2006, they 'predicted that a low 2D:4D would be associated with high levels of egoism and altruism and low levels of common cooperativeness (i.e. contributing exactly one's fair share).' However, they 'found the exact opposite'.

There are hundreds of other finger celebrities. Dr Mark Brosnan, a psychology researcher at the University of Bath, may be the most notorious. Brosnan studied the fingers of one hundred of his colleagues at the university. He reports that the digit ratio is significantly different for science teachers than for those who teach humanities or social science. No one has tried to dispute this finding.

Phelps, V. Rae (1952). 'Relative Index Finger Length as a Sex-influenced Trait in Man'. *American Journal of Human Genetics* 4 (2): 72–89.

Bennett, M., John T. Manning, Christian J. Cook and Liam P. Kilduff (2010). 'Digit Ratio (2D:4D) and Performance in Elite Rugby Players'. *Journal of Sports Sciences* 28 (13): 1415–21.

Manning, John T. (2008). *The Finger Book*. London: Faber and Faber.

Nelson, Emma, John T. Manning and Anthony G.M. Sinclair (2006). 'Using the 2nd and 4th Digit Ratio (2D:4D) to Sex Cave Art Hand Stencils: Factors to Consider'. *Before Farming* 1: n.p.

Nelson, Emma, Christy L. Hoffman, Melissa S. Gerald and Susanne Shultz (2010). 'Finger Length Ratios (2D:4D) and Dominance Rank in Female Rhesus Macaques (*Macaca mulatta*)'. *Behavioral Ecology and Sociobiology* 64: 1001–9.

Nelson, Emma, Campbell Rolian, Lisa Cashmore and Susanne Shultz (2011). 'Digit Ratios Predict Polygyny in Early Apes, *Ardipithecus,* Neanderthals and early Modern Humans But Not in *Australopithecus*'. *Proceedings of the Royal Society B* 278: 1556–63.

Voracek, Martin, John T. Manning and Ivo Ponocny (2005). 'Digit Ratio (2D:4D) in Homosexual and Heterosexual Men from Austria'. *Archives of Sexual Behavior* 34 (3): 335–40.

Voracek, Martin, and John T. Manning (2003). 'Length of Fingers and Penis Are Related through Fetal *Hox* Gene Experession'. *Urology* 62 (1): 201.

Voracek, Martin, Barbara Reimer, Clara Ertl and Stefan G. Dressler (2006). 'Digit Ratio (2D:4D), Lateral Preferences, and Performance in Fencing'. *Perception and Motor Skills* 103 (2): 427–46.

Voracek, Martin, Barbara Reimer and Stefan G. Dressler (2010). 'Digit Ratio (2D:4D) Predicts Sporting Success among Female Fencers Independent from Physical, Experience, and Personality Factors'. *Scandinavian Journal of Medicine & Science in Sports* 20 (6): 853–60.

Kilduff, Liam P., Christian J. Cook and John T. Manning (2011). 'Digit Ratio (2D:4D) and Performance in Male Surfers'. *Journal of Strength and Conditioning Research* 25 (11): 3175–80.

Williams, Terrance J., Michelle E. Pepitone, Scott E. Christensen, Bradley M. Cooke, Andrew D. Huberman, Nicholas J. Breedlove, Tessa J. Breedlove, Cynthia L. Jordan and S. Marc Breedlove (2000). 'Finger-length Ratios and Sexual Orientation'. *Nature* 404 (30 March): 455–6.

Brown, Windy M., Christopher J. Finn, Bradley M. Cooke, and S. Marc Breedlove (2002). 'Differences in Finger Length Ratios Between Self-identified "Butch" and "Femme" Lesbians'. *Archives of Sexual Behavior* 31 (1): 123–7.

Brosnan, Mark J. (2006). 'Digit Radio and Faculty Membership: Implications for the Relationship between Prenatal Testosterone and Academia'. *British Journal of Psychology* 97: 455–66.

— (2008). 'Digit Ratio as an Indicator of Numeracy Relative to Literacy in 7-year-old British Schoolchildren'. *British Journal of Psychology* 99: 75–85.

—, V. Galllop, N. Iftikhar and E. Keogh (2011). 'Digit Ratio (2D:4D), Academic Performance in Computer Science and Computer-related Anxiety'. *Personality and Individual Differences* 51 (4): 371–5.

Voracek, Martin, John T. Manning and Stefan G. Dressler (2007). 'Repeatability and Interobserver Error of Digit Ratio (2D:4D) Measurements Made by Experts'. *American Journal of Human Biology* 19 (1): 142–6.

Millet, Kobe and Siegfried Dewitte (2006). 'Second to Fourth Digit Ratio and Cooperative Behavior'. *Biological Psychology* 71 (1): 111–15.

WHAT AM I MISSING?

Fuelled with curiosity, some scientists exploit – lovingly, proudly – the investigative trick featured in Arthur Conan Doyle's 1892 story 'Silver Blaze'. There, a baffled police inspector seeks help from the great autodicact/detective Sherlock Holmes:

> [INSPECTOR GREGORY:] 'Is there any point to which you would wish to draw my attention?'
>
> [HOLMES:] 'To the curious incident of the dog in the night-time.'
>
> [INSPECTOR GREGORY:] 'The dog did nothing in the night-time.'
>
> 'That was the curious incident', remarked Sherlock Holmes.

Science journals feature many papers in which scientists rely on this technique, riding it to, or at least in the direction of, glory. You can see that happening in a report called 'The Mystery

of the Missing Toes: Extreme Levels of Natural Mutilation in
Island Lizard Populations', published in 2009 in the journal
Functional Ecology.

The co-authors – Bart Vervust, Stefan Van Dongen and
Raoul Van Damme at the University of Antwerp, Belgium,
and Irena Grbac at the Natural History Museum of Croatia –
'report on an exceptionally large difference in toe-loss incidence
between two populations of *Podarcis sicula* lizards living on
small, neighbouring islands in the Adriatic Sea. We caught 900
lizards and recorded the number and location of missing toes.'
Having gathered that data, the scientists then walked through
the logic of 'five non-mutually exclusive hypotheses concerning
differences in bite-force capacity, bone strength ...' and so forth.

Unlike the fictional British detective, this very real Belgian/
Croatian research team failed to discover a tidy, satisfying solution to their mystery. Nonetheless, they found reason for cheer,
explaining that 'such tests can reveal how likely each of these
explanations is, even if the processes leading to the phenomenon
are difficult to observe directly.'

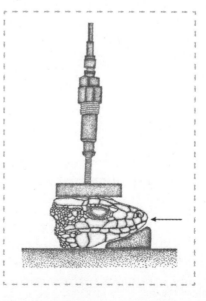

Figure: 'Device for measuring bite force needed for breaking a toe' from 'The Mystery of the Missing Toes'

The method sometimes flops. That's evident in a different study about lizards, published by the Chilean/US team of Fabian Jaksic and Stephen Busack in 1984 in the journal *Amphibia-Reptilia*. Jaksic and Busack sum things up in their title: 'Apparent Inadequacy of Tail-loss Figures as Estimates of Predation upon Lizards'.

Some scientists spurn or ignore the method, or find that it does not apply to their particular investigation, however. In 1994, G.J. Adams and K.G. Johnson at Murdoch University in Australia published a study with what appears to be a blatantly, proudly Sherlock Holmesian title. Adams and Johnson called their report 'Behavioural Responses to Barking and Other Auditory Stimuli during Night-time Sleeping and Waking in the Domestic Dog (*Canis familiaris*)'. Curiously, Adams and Johnson neither use nor allude to the Sherlock Holmes trick.

They explain that they filmed twelve dogs 'at night in their usual urban habitats, whilst alert, in quiet sleep and in active sleep'. They subjected each dog to six different audio recordings: 'a single bark; repeated barking; breaking glass; a motorcycle; a bus; and 'rowdy young people discussing burglurizing [sic]'. They discovered, they say, that 'dogs were found to be significantly more responsive to auditory stimuli when alert than when asleep.'

Vervust, Bart, Stefan Van Dongen, Irena Grbac, and Raoul Van Damme (2009). 'The Mystery of the Missing Toes: Extreme Levels of Natural Mutilation in Island Lizard Populations'. *Functional Ecology* 23 (5): 996–1003.

Jaksic, Fabian M., and Stephen D. Busack (1984). 'Apparent Inadequacy of Tail-loss Figures as Estimates of Predation upon Lizards'. *Amphibia-Reptilia* 5: 177–9.

Adams, G.J., and K.G. Johnson (1994). 'Behavioural Responses to Barking and Other Auditory Stimuli during Night-time Sleeping and Waking in the Domestic Dog (*Canis familiaris*)'. *Applied Animal Behaviour Science* 39 (2): 151–62.

RESEARCH SPOTLIGHT

'RELATIONSHIPS OF TOE-LENGTH RATIOS TO FINGER-LENGTH RATIOS, FOOT PREFERENCE, AND WEARING OF TOE RINGS'

by Martin Voracek and Stefan G. Dressler (published in *Perceptual and Motor Skills*, 2010)

ON PEG-LEG COYOTES

'Running Speeds of Crippled Coyotes' introduced itself in 1976, in a journal called *Northwest Science*. You'll find few scientific studies that tell their story so clearly and efficiently. Bruce C. Thompson, of the department of fisheries and wildlife at Oregon State University, wrote everything he had to say in a plain two pages.

It contains little jargon or lingo, and no clever metaphors. When the study speaks of crippled coyotes, it means exactly that: coyotes that are crippled.

Thompson begins with some history, just enough so you learn that other people, in earlier days, spent time thinking about how fast coyotes run. He alludes to decades-old studies, by scientists named Cottam, Sooter and Zimmerman, that 'reported running speeds of presumably uninjured coyotes being chased by cars'.

Thompson brought something new to the table (so to speak): 'On 21, 22, and 23 October 1974, I recorded running speeds of three wild-trapped coyotes that had lost the use of one foot due to damage from a steel trap.'

How exactly did Thompson accomplish this? He tells you in just a few sentences: 'During the tests, the coyotes were released from their cages singly and allowed to run along the perimeter fence of the enclosure. Each day the coyotes were timed with a stop-watch as they ran three measured courses along the perimeter fence. As a coyote approached the starting point of each course, I chased the animal on foot at a distance of 45 meters to 70 meters.'

Thompson also measured the running speed of a coyote that had all its original equipment. On its best run, that animal had a speed of just under thirty-two miles per hour. One of the crippled animals matched that almost exactly, despite lacking a right foot. The other three-footed coyotes attained best speeds of 22.5 miles per hour and 25.4 miles per hour, respectively. (The full-bodied coyotes chased by four-wheeled cars decades earlier, by the way, ran much faster than the one chased in the 1970s by the two-legged Bruce Thompson.)

Thompson also paid attention to style. 'Although the crippled coyotes occasionally contacted the ground with their damaged appendage', he wrote, 'they typically adjusted their stride to prevent contact with the ground. The adjusted stride resulted in a noticeable bouncing movement when the crippled coyotes ran.'

Bruce Thompson's monograph refers, glancingly, to a 1939 study called 'Food Habits of Peg-leg Coyotes', by Charles C. Sperry of the US Biological Survey's Food Habits Laboratory in Colorado. Sperry, too, knew how to tell a tale. Who could resist this beginning? 'During the past two years, 164 peg-leg coyote stomachs that contained food remains were obtained and their contents examined in the Denver laboratory.'

I will skip over Sperry's other good parts, and get right to his thrilling conclusion: 'It will be noted that two peg-leg coyotes eat as much livestock as three normal coyotes.

Thompson, Bruce C. (1976). 'Running Speeds of Crippled Coyotes'. *Northwest Science* 50 (3): 181–2.
Cottam, Clarence (1945). 'Speed and Endurance of the Coyote'. *Journal of Mammalogy* 26 (1): 94.
—, and C.S. Williams (1943). 'Speed of Some Wild Mammals'. *Journal of Mammalogy* 24 (2): 262–3.
Sooter, C.A. (1943). 'Speed of Predator and Prey'. *Journal of Mammalogy* 24 (1): 102–3.
Zimmerman, R.S. (1943). 'A Coyote's Speed and Endurance'. *Journal of Mammalogy* 24 (3): 400.
Sperry, Charles C. (1939). 'Food Habits of Peg-leg Coyotes'. *Journal of Mammalogy* 20 (2): 190–4.
Young, S.P., and H.H.T. Jackson (1951). *The Clever Coyote*. Harrisburg, PA, and Washington, DC: Stackpole Co. and Wildlife Management Institute.

AN IMPROBABLE INNOVATION

'ANIMAL TRACK FOOTWEAR SOLES'

a/k/a shoes for laying stimulated animal tracks 'for either educational purposes or mere amusement', by Philip E. McMorrow (US patent no. 3,402,485, granted 1968)

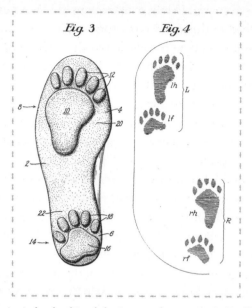

The patented animal track footwear soles, compared to Kodiak bear cub tracks

PUMPKIN HARVEST

The knifing of pumpkins, an innocent-seeming yet carefully planned act of mutilation, sometimes results (accidentally or otherwise) in sprays, bits and smatterings of human, as well as vegetable, gore. In such cases, blood – human blood – flows, drips and coagulates.

A hands-on experiment, or rather an experiment on hands, in 2004, tried to determine the level of medical danger an amateur can and should expect when using a pumpkin-carving tool.

Alexander M. Marcus, Jason K. Green and Frederick W. Werner at the State University of New York Upstate Medical University in Syracuse published a study, called 'The Safety of Pumpkin-Carving Tools', in the journal *Preventive Medicine*. 'Pumpkin-carving accidents', they inform their peers who read the report, 'may leave people with compromised hand function'.

There are several kinds of lacerations and puncture wounds that can lead to this hand-compromisation. Lacerations occur 'when the knife blade travels across the surface of the hand' or 'when the knife is accidentally pushed too far forward and cuts the opposite hand stabilising the pumpkin', or 'when the cutting hand slips forward off the handle and on to the blade', in which case 'injury occurs across zone 2 of the volar surface [the palm] of the hand, while the flexor tendons are taut from gripping the knife'.

Marcus, Green and Werner take pains to educate those among their fellow physicians who may lack experience in recognizing and treating puncture wounds. 'Puncture wounds', they write, 'occur when the point of the knife contacts the hand while traveling perpendicular to it.'

Their experiments compared two commercially offered pumpkin-cutting knives (the Pumpkin Kutter versus the Pumpkin Masters' Medium Saw) and two ordinary kitchen knives, one serrated, one plain-edged.

They measured how much force was required to make a cut.

The first experiment used a machine to plunge each knife into a pumpkin, an actuator driving the weapon downward at a rate of three millimetres per second, each knife being inserted at four different penetration sites on each of three pumpkins, testing with the blade being oriented along the grooves of the pumpkin, and separately with it perpendicular to those grooves. One of the doctors also tested what happened when he, not a machine, attacked the pumpkins with a sawing motion. The pumpkin knives required 'statistically' less force than the kitchen knives to cut pumpkins.

In keeping with the festive theme of the activity, the physicians used 'a cadaver model' in their second experiment. They obtained 'six cadaver forearms ... harvested at the elbow'. The report includes photographs showing how this played out.

The results: First, it took less force to cut a person than a pumpkin. Second, the kitchen knives required statistically less force to penetrate cadaver skin, and 'caused more skin lacerations that would require suturing than either pumpkin knife'.

'Experimental setup to cause tendon laceration in cadaver hands'

Marcus, Alexander M., Jason K. Green and Frederick W. Werner (2004). 'The Safety of Pumpkin Carving Tools'. *Preventive Medicine* 28 (6): 799–803.

IN BRIEF

'PANTI-GIRDLE SYNDROME'
by T.K. Davidson (published in the *British Medical Journal*, 1972)

'THE TIGHT-GIRDLE SYNDROME'
by P.D. White (published in the *New England Journal of Medicine*, 1973)

'PANTY HOSE-PANTS DISEASE'
by M. Turner (published in the *American Journal of Obstetrics and Gynecology*, 1991)

'TIGHT PANTS SYNDROME: A NEW TITLE FOR AN OLD PROBLEM AND OFTEN ENCOUNTERED MEDICAL PROBLEM'
by Octavio Bessa Jr (published in the *Archives of Internal Medicine*, 1993)

CUTTING-EDGE ENGINEERING

Through the clever use of cheese in 2004, researchers at the University of Reading claimed to have solved one of life's great little mysteries. 'Why is it relatively difficult, even with a sharp knife, to cut when simply "pressing down", but much easier to cut as soon as some sideways sawing or slicing action is introduced?'

The scientists, A.G. ('Tony') Atkins, George Jeronimidis and X. Xu, published a monograph called 'Cutting, by "Pressing and Slicing" of Thin Floppy Slices of Materials Illustrated by Experiments on Cheddar Cheese and Salami'. It was the highlight of that April's issue of the *Journal of Materials Science*.

The team experimented on a piece of cheese that they identified only as 'commercial Cheddar cheese'.

They were equally cagey about the nature of the meat. The report employs the phrase 'a commercial pepper salami'. Some salami experts understand that those four words, when huddled together, cover a multitude of possibilities: delicious or not, cheering or horrifying, mushy, stiff or adamantine. They and we learn nothing about this particular salami, save that thin slices of it become 'floppy'. But that's all right, because floppiness is the key thing here.

Atkins, Jeronimidis and Xu write in fairly stiff, technical language. But they loosen up a bit when talking about floppiness. Then they use language suitable to the casual reader who might, say, peruse an issue of the *Journal of Materials Science* while lounging in a dentist's waiting room: 'Further examples of the sort of globally-elastic cutting considered in this paper', say the scientists, 'are to be found in the slicing of meat by the butcher ... lawnmowing, hair cutting, the cutting of fabrics by the dressmaker, surgery and so on. These cases are characterised by the offcuts [the individual slices] being elastically very floppy (ie, have negligible bending resistance and are not permanently deformed).'

Cutting into non-floppy material can be a different game. Atkins, Jeronimidis and Xu explicitly leave that for others to investigate.

They chose cheese, they say, because 'the cutting of cheese is notoriously affected by friction (hence the use of wire to cut cheese)'. They do not explain why they selected salami.

The team did their cutting not with a wire, but with a delicatessen-style 'bacon' slicer. Its whirling blade can, depending on the substance and on the angle of the cut, fall prey to varying amounts of friction. To keep things from getting too, too floppy, they chilled the cheese before slicing it, and ditto the salami.

They quantified, in unprecedentedly technical detail, what good butchers, lawnmowers, hair cutters, dressmakers and surgeons have always intuited. The faster the whirl of the blade (or the horizontal drawing of the knife), the less force is needed to drive the blade down, down, down into the material. But, because of friction – the rubbing of blade against substance – there's a limit to how easy that downward slicing can get. Cutting-edge stuff.

Atkins, A.G. ('Tony'), X. Xu and George Jeronimidis (2004). 'Cutting, by "Pressing and Slicing" of Thin Floppy Slices of Materials Illustrated by Experiments on Cheddar Cheese and Salami'. *Journal of Materials Science* 39 (8): 2761–6.

IN BRIEF

' "YOUR FEET'S TOO BIG": AN INQUIRY INTO PSYCHOLOGICAL AND SYMBOLIC MEANINGS OF THE FOOT'
by K.J. Zerbe (published in *Psychoanalytic Review*, 1985)

THE SECRET WEALTH IN A COCKROACH LEG

Biscuits, rubbish and bugs in Texas raise hopes that Britain will grow a lucrative new technology-based empire soon, rather than just eventually. This is all about getting usable amounts of graphene – the two-dimensional form of carbon. An American experiment, so goofy-sounding that it has drawn little attention, points towards a cheap way of obtaining what is now a scarily expensive substance.

Scientists had long known that graphene exists, and that it is common. The grey stuff in pencils is made of multitudinous layers of graphene, sticking to each other. When you scribble, a gob of layers slides away, clinging thereafter to your sheet of paper. A few years ago, physicists Andre Geim and Kostya Novoselov, at the University of Manchester, used cleverness, a pencil and sticky tape to separate out some single layers of graphene. They obtained only tiny amounts – but that was staggeringly more than anyone else had managed.

For doing that, and for then using their graphene to discover a multitude of physical properties and likely industrial uses, Geim and Novoselov were given a Nobel Prize in 2010, and knighthoods in the 2012 New Year Honours. Later that year the British government announced it would spend £38 million to establish a Geim/Novoselov-centric National Graphene Institute at Manchester University, aimed at 'taking this research through to commercial success'.

But there is a big problem. Even tiny amounts of graphene still cost far more than industry can dream of affording.

Enter the North Americans.

In 2011, just three years after some Mexican scientists converted tequila into diamonds – which are just an expensive form of carbon – chemists in Texas quietly mucked around with some cookies, cockroaches and disgusting biological waste products. The Texans produced graphene, a form of carbon dearer – much dearer – than diamonds.

The tequila-into-diamonds physicists, at the Universidad Nacional Autónoma de México, published their story in a monograph called 'Growth of Diamond Films from Tequila'. Co-author Javier Morales said they later made diamonds using the cheapest tequila on the market, to demonstrate their technique's power.

That same spirit is evident in the Texas experiment, performed by James M. Tour, Gedeng Ruan, Zhengzong Sun and Zhiwei Peng at Rice University in Houston, and documented in a study called 'Growth of Graphene from Food, Insects, and Waste'.

Others, elsewhere, had devised ways to grow graphene. Those other methods begin with costly, highly purified carbon-containing chemical feedstocks. Tour and his Rice colleagues write that 'much less expensive carbon sources, such as food, insects and waste, can be used without purification to grow high-quality monolayer graphene'.

They explain how they produced graphene from (at different times) Girl Scout cookies, chocolate, grass, a plastic dish, a cockroach leg and faeces from a dachshund. Their tools: copper foil, argon gas and a 1050-degree Celsius tube furnace.

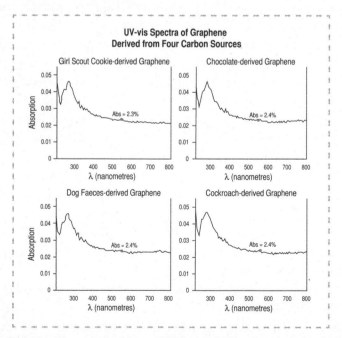

Analysing graphene derived from a Girl Scout cookie, chocolate, dog faeces and a cockroach

By their estimate, one box of Girl Scout cookies could theoretically, at then-current prices, be converted to about $15 billion worth of graphene.

Morales, Apátiga and Casataño were awarded the 2009 Ig Nobel Prize in chemistry for creating their tequila-derived diamonds.

Novoselov, Kostya, S., Andre K. Geim, Sergey V. Morozov, Da Jiang, Yuanbo Zhang, Sergey V. Dubonos, Irina V. Grigorieva and Anatoly A. Firsov (2004). 'Electric Field Effect in Atomically Thin Carbon Films'. *Science* 306 (5696): 666–9.

Novoselov, Kostya, S., Da Jiang, Frederick Schedin, Tim J. Booth, V.V., Khotkevich, Sergey V. Morozov and Andre K. Geim (2005). 'Two-dimensional Atomic Crystals'. *Proceedings of the National Academy of Sciences of the USA* 102 (30): 10451–3.

Morales, Javier, Miguel Apátiga and Victor M. Casataño (2009). 'Growth of Diamond Films from Tequila'. *Reviews on Advanced Materials Science* 22 (1): 134–8.

Ruan, Gedeng, Zhengzong Sun, Zhiwei Peng and James M. Tour (2011). 'Growth of Graphene from Food, Insects, and Waste'. *ACS Nano* 5 (9): 7601–7.

IN BRIEF

'TESTING THE VALIDITY OF THE DANISH URBAN MYTH THAT ALCOHOL CAN BE ABSORBED THROUGH FEET: OPEN LABELLED SELF EXPERIMENTAL STUDY'

by Christian Stevns Hansen, Louise Holmsgaard Faerch and Peter Lommer Kristensen (published in the *British Medical Journal,* 2010)

The authors, at Hillerod Hospital, Hillerod, Denmark, explain: 'Objective: To determine the validity of the Danish urban myth that it is possible to get drunk by submerging feet in alcohol. Participants: Three adults, median age 32 (range 31-35) … Conclusion: Our results suggest that feet are impenetrable to the alcohol component of vodka.'

SIX

NAVEL GAZING, CURIOUS CONSUMING

IN BRIEF

'BITING OFF MORE THAN YOU CAN CHEW: A FORENSIC CASE REPORT'
by J.R. Drummond and G.S. McKay (published in the *British Dental Journal*, 1999)

Some of what's in this chapter: Query your belly • Synchronize your cows • Know thy fly • Colour-change thy cereal • Drink in your results • Meet your meat, in unintended circumstances • Peruse your hot potatoes • Disgust your carnivore shrink • Define your vegetarian, strictly • Weigh your falafel • Identify your water • Sneak-peek your hosts' food • Slim up your fatter fellows • Chew your crisp cereal • Cereal-flake your cows • Scrawl your boozy scrawl • Note your neighbours in the pub

EXPOSING THE GERMAN BEER BELLY

A team of scientists has attacked the idea that beer is the main cause of beer bellies in Germans. As with many biomedical questions, an absolute, indisputable answer may be impossible. To obtain it would require continuously monitoring and measuring, over a span of years, every sip and morsel drunk and eaten by a vast number of people.

The only practical method involves asking beer-drinkers to

dig deep into their memory and estimate or take a wild guess as to their typical intake of beer and everything else that has passed down their gullet.

Madlen Schütze and other researchers at the German Institute of Human Nutrition Potsdam-Rehbrücke and at Fulda University of Applied Sciences, together with a colleague at the University of Gothenburg, Sweden, gave it their best shot in the *European Journal of Clinical Nutrition*.

The team analysed 19,941 men and women's weight, waist measurements and hip circumference over four years. The researchers asked participants to fill out a survey on their beer consumption.

The participants gave an estimate, with whatever degree of accuracy they were able or willing to supply, of how much they had been imbibing daily. Their estimates were based on the size of a typical bottle of beer in Germany.

The researchers classified men's consumption differently to women's. Women were placed in four categories from 'no beer' to 'moderate drinkers'. Men were put in five categories from 'no beer' to 'heavy drinkers'. For women, 'moderate' meant consuming at least 250 millilitres of beer a day. For men, 'moderate' meant 500 to 1,000 millilitres a day.

The team conclude that their study 'does not support the common belief of a site-specific effect of beer on the abdomen, the beer belly'. 'Beer consumption', they write, 'seems to be rather associated with an increase in overall body fatness'.

A study in the Czech Republic, published several years earlier in the same medical journal, balks at the idea that drinking beer by itself causes much change in weight, let alone waistlines, in Czechs. The study aimed to investigate the 'common notion that beer drinkers are, on average, more "obese" than either nondrinkers or drinkers of wine or spirits'. Schütze and colleagues argue that that Czech study was probably flawed, that its findings were a bit bloated.

One can quibble about the definition of a beer belly, but the German researchers say that, for their purposes, a beer belly is a 'site-specific effect'. A beer belly, they argue, is a belly that bulges distinctly at the waist. It contrasts, in a big way, with whatever mass and expanse may adjoin it above or below that region.

Schütze, Madlen, Mandy Schulz, Annika Steffen, Manuela M. Bergmann, Anja Kroke, Lauren Lissner and Heiner Boeing (2009). 'Beer Consumption and the 'Beer Belly': Scientific Basis or Common Belief?'. *European Journal of Clinical Nutrition* 63 (9): 1143–9.
Bobak, Martin, Zdenka Skodova and Michael Marmot, (2003).'Beer and Obesity: A Cross-sectional Study'. *European Journal of Clinical Nutrition* 57 (10): 1250–3.

SYNCHRONIZED COWS

A British-American team of scientists has produced a study called 'A Mathematical Model for the Dynamics and Synchronization of Cows'. They were driven partly by the intellectual challenge, and at least a little by an EU council directive (precisely, council directive 97/2/EC), which mandates 'that cattle housed in groups should be given sufficient space so that they can all lie down simultaneously'.

Their key insight, the team says, was to realize 'it is biologically plausible to view [cattle] as oscillators ... During the first stage (standing/feeding), they stand up to graze but they strongly prefer to lie down and "ruminate" or chew the cud for the second stage (lying/ruminating). They thus oscillate between two stages.'

The researchers 'modeled the eating, lying, and standing dynamics of a cow using a piecewise linear dynamical system ... We chose a form of coupling based on cows having an increased desire to eat if they notice another cow eating and an increased desire to lie down if they notice another cow lying down.' This, they say, led to at least one unexpected discovery: '[We] showed that it is possible for cows to synchronize *less* when the coupling is increased.'

The researchers – Mason Porter and Marian Dawkins at Oxford University, and Jie Sun and Erik Bollt at Clarkson University in Potsdam, New York – published their work in the physics journal *Physica D: Nonlinear Phenomena*. In the thirty-one-year history of that journal, this was the first article specifically about cows. (Cows do have an accepted, very humble place in the history of physics: an old joke, beloved by physicists and a few others. The joke starts [usually] with a physicist offering to solve a dairy-related problem for a desperate farmer. The physicist walks to a blackboard, and draws a circle. 'First', he says, 'we assume a spherical cow …').

We model the dynamics of a single cow in different states using

$(\mathscr{E})$ Eating state: $\begin{cases} \dot{x} = -\alpha_2 x, \\ \dot{y} = \beta_1 y. \end{cases}$ (3)

$(\mathscr{R})$ Resting state: $\begin{cases} \dot{x} = \alpha_1 x, \\ \dot{y} = -\beta_2 y. \end{cases}$ (4)

$(\mathscr{S})$ Standing state: $\begin{cases} \dot{x} = \alpha_1 x, \\ \dot{y} = \beta_1 y. \end{cases}$ (5)

where the calligraphic letters inside parentheses indicate the corresponding values of θ. For biological reasons, the parameters $\alpha_1, \alpha_2, \beta_1,$ and β_2 must all be positive real numbers. They can be interpreted as follows:

$\begin{cases} \alpha_1 : \text{rate of increase of hunger,} \\ \alpha_2 : \text{decay rate of hunger,} \\ \beta_1 : \text{rate of increase of desire to lie down,} \\ \beta_2 : \text{decay rate of desire to lie down.} \end{cases}$

Decoding a cow's hunger versus its desire to lie down, mathematically

The team built upon the work of earlier, fully serious bovimathematical scholars.

In 1982, P.F.J. Benham of Reading University published an innovative study called 'Synchronization of Behaviour in Grazing Cattle'. Brennan studied a herd of thirty-one Friesian cows, recording the behaviour of each every five minutes during daylight for fifteen days. His short paper – it's only two pages long – ends with the declaration: 'Studies of behaviour synchronization are evidently relevant to the management of grazing cattle.'

Porter, Dawkins, Sun and Bollt also looked beyond the bounds of cow analysis, gaining insight from a 1991 monograph by A.J. Rook and P.D. Penning of the AFRC Institute of Grassland and Environmental Research in Hurley, Maidenhead. Rook and

Penning called their report 'Synchronisation of Eating, Ruminating and Idling Activity by Grazing Sheep', and published it in the journal *Applied Animal Behaviour Science*. They reached four conclusions. I will mention only one, as it has wide applicability: 'Start of meals was more synchronised than end of meals.'

Sun, Jie, Erik M. Bollt, Mason A. Porter and Marian S. Dawkins (2011). 'A Mathematical Model for the Dynamics and Synchronization of Cows'. *Physica D: Nonlinear Phenomena* 240: 1497–1509.

Benham, P.F.J. (1982). 'Synchronization of Behaviour in Grazing Cattle'. *Applied Animal Ethology* 8 (4): 403–4.

Rook, A.J., and P.D. Penning (1991). 'Synchronisation of Eating, Ruminating and Idling Activity by Grazing Sheep'. *Applied Animal Behaviour Science* 32 (2/3): 157–66.

TO KNOW A FLY

Vincent Dethier loved flies with a fervour that is rare. He distilled this love into a book called *To Know a Fly*.

Page fifty-three tells what happens when Dethier severed the tiny nerve that tells a fly whether it has had enough to eat: 'The results of this operation on a hungry fly were spectacular. Such a fly began to eat in the normal fashion, but did it stop? Never. It ate and ate and ate. It grew larger and larger. Its abdomen became so stretched that all the organs were flattened against the sides. It became so big and round and transparent that it could almost be used as a miniature hand lens. It was so round its feet no longer reached the ground and so heavy it could not launch itself into the air. Even though the back pressure from a near bursting crop was terrific, the fly continued in its attempts to eat. It reminded me of a woman who had been admitted to our hospital, a woman whose height was four feet, ten inches and whose weight approached four hundred pounds. Her major complaint was inability to move.'

In just 119 pages Dethier describes many of the fly's unadvertised charms and wonders. He makes no pretence of giving explanations for particular wonders that neither he nor any other scientist really understands. This in itself is wonderful and charming. Consider: 'We know ... there is a time when the female fly prefers protein, which cannot nourish her own body, to sugar, which is an adequate food for her but useless for her eggs. Here is an example of survival of the individual being subordinated to survival of the species. In some quarters it would be hailed as maternal instinct, and by so naming it we would be no nearer an understanding of what it is.'

Dethier hungered still for information about flies, producing many detailed studies, including a 489-page book, *The Hungry Fly*, in 1976.

To Know a Fly first appeared in 1962. Dethier was a biologist based at Princeton and, at various times, at other universities. There is poetry in his book, but not the lugubrious kind that makes practical people flee. Many chapters begin with brief passages from Don Marquis's 1927 book *Archy and Mehitabel*, which is the source of much modern wisdom about cockroaches and cats (and perhaps about people, too). Had Dethier's book appeared first, it would not have been out of place as source of chapter lead-in material for Marquis and Archy.

I learned about *To Know a Fly* from Shelly Marino of Cornell University, who described it as 'the book that turned me into a biologist in the first place'. This book can do for flies what the Harry Potter films have done for Daniel Radcliffe and Emma Watson. It is magically powerful stuff.

Dethier, Vincent G. (1962). *To Know a Fly*. San Francisco: Holden-Day.
— (1976). *The Hungry Fly: A Physiological Study of the Behavior Associated with Feeding*. Cambridge: Harvard University Press.
Marquis, Don (1927). *Archy and Mehitabel*. Garden City, NY: Doubleday, Page.

IN BRIEF

'SPECIALIST ANT-EATING SPIDERS SELECTIVELY FEED ON DIFFERENT BODY PARTS TO BALANCE NUTRIENT INTAKE'
by S. Pekár, D. Mayntz, T. Ribeiro and M.E. Herberstein (published in *Animal Behaviour*, 2010)

CHAMELEON CRUNCH

When parents warn children not to play with their food, there's now reason to add a menacing 'even if': 'even if the food begins playing with you'. Recently, food was given a new ability to play, a little, the moment it encounters milk.

Researchers have patented a way to make breakfast cereal change colour as it sits in the bowl, awaiting its roller-coaster ride down somebody's throat.

The patent documents explain why the world needs this to occur, as well as how, chemically and mechanically, to do it.

Hideo Tomomatsu of Crystal Lake, Illinois, filed a patent application in 1987 for what he called 'colour-changing cereals'. Eight years later, Joseph Farinella of Chicago, Illinois, and Justin French of Cedars, Iowa, used much of the same stilted wording in filing their own application. Both patents were granted, with the rights being assigned to the Quaker Oats Company. Quaker's colour-changing was in 'Cap'n Crunch' cereal – expressly, the 'Polar Crunch' variety, which apparently was on sale as early as 2006.

Powder Coating Composition

Sugar (parts by weight)	Starch (parts by weight)	Color (parts by weight)	Time to complete color change (seconds)
85	14	1	25
34	65	1	15
24	75	1	10
24	99	1	7

Particular coating compositions tested. The inventors note: 'While the invention has been described with respect to certain preferred embodiments, as will be appreciated by those skilled in the art, it is to be understood that the invention is capable of numerous changes, modifications and rearrangements' which, they say, 'are intended to be covered' by the patent.

The Quaker Oats Company, founded in 1901, makes breakfast cereal – buckets and buckets of it. Playing with food is good for its industry. Quaker even partially financed the apotheosis of that activity: the 1971 film version of Roald Dahl's book *Charlie and the Chocolate Factory*.

The patent language puts the new, shade-shifting food in context: 'One approach to engage the interest of children, as well as other people, in eating RTE [Ready to Eat] cereal would be an RTE cereal that changes colour on contact with an aqueous edible medium such as cold milk, hot water, etc. To make the change more interesting, the colour change should be rapid and substantial.'

'There is a need' for this, the inventors write.

'There is also a need', they continue, 'for a method for making colour-changing cereals that is efficient and cost-effective.'

Their method is to create cereal pieces of one colour, then coat them with powder of a different hue. That leads to breakfast table magic: 'The coating is of a second colour different from the first colour and is in a quantity sufficient to obscure the first colour … Upon mixing milk with the resulting cereal, the edible powdered surface is instantly dissolved or dispersed, revealing the specific colours of the individual pieces very quickly.'

What are the ingredients? Glad you asked. 'The cereal base has a coating comprising cornstarch, powdered sugar, [and] food colouring.'

The researchers tinkered with the recipe, to see how quickly they could make the cereal disrobe. That resulted in what they believe to be a scientific discovery: 'Surprisingly, the use of cornstarch in the correct ratio to powdered sugar increases the speed of the colour change. This creates a more startling effect that is appealing to children.'

That ratio, with a cereal coating of mostly starch and just a smidge of sugar, got the transformation down to a presto-change-o, what-a-way-to-start-the-day seven seconds.

Tomomatsu, Hideo (1989). 'Color-changing Cereals and Confections'. US patent no. 4,853,235, 1 August.
Farinella, Joseph R., and Justin A. French (2010). 'Color-changing Cereal and Method'. US patent no. 7,648,722, 19 January.

DRINKING IN THE RESULTS

America, a rich source of alcohol, of alcoholics, and of aggressive alcoholics, is also rich in scholarship on those subjects. One must drink deep of that scholarship, in many cases, if one cares about the question: what, exactly, did some of those researchers hope to learn by doing that research? The flow of prose produced by these researchers could drown anyone who tried to ingest more than occasional, measured amounts of it.

In a 1999 study called 'The Effects of a Cumulative Alcohol Dosing Procedure on Laboratory Aggression in Women and Men', Donald Dougherty and colleagues at the University of Texas-Houston medical school report making several discoveries: (1) Both men and women became more aggressive after they drank alcohol; (2) those men and women became even more aggressive 'after consuming the second alcoholic drink'; (3) their aggressiveness 'remained elevated for several hours' after they finished drinking; and (4) the individuals who were most aggressive when sober were also the most aggressive when drunk.

Three years later, Peter Giancola at the University of Kentucky wrote a study called 'Irritability, Acute Alcohol Consumption and Aggressive Behaviour in Men and Women'. 'The finding of greatest importance', Giancola says, 'was that alcohol increased aggression for persons with higher, as opposed to lower, levels of irritability.'

The following year, Dominic J. Parrott, Amos Zeichner and Dana Stephens at the University of Georgia published a treatise called 'Effects of Alcohol, Personality, and Provocation on the Expression of Anger in Men: A Facial Coding Analysis'.

Parrott and his colleagues experimented, and made two big discoveries. First: that drunken people made 'facial expressions of anger' more often than sober people. Second: that when highly provoked, drunks also showed a general 'tendency to express anger outwardly'.

Many other alcohol mysteries are explained in these same research journals. In 1995, Siegfried Streufert and fellow researchers at Penn State boozed up some managers. Streufert published a study, called 'Alcohol Hangover and Managerial Effectiveness'. It reports that managers who drank moderately of an evening had perfectly adequate 'complex decision-making competence' the next morning.

Suzanne Thomas, Carrie Randall and Maureen Carrigan at the Medical University of South Carolina wrote a paper in 2003, called 'Drinking to Cope in Socially Anxious Individuals: A Controlled Study'. They explain: 'The results of this study confirm earlier observations that individuals high in social anxiety deliberately drink to cope with social fears'.

In 2005, Soyeon Shim at the University of Arizona and Jennifer Maggs at Penn State University went looking for results. They found some. Their study called 'A Cognitive and Behavioral Hierarchical Decision-making Model of College Students' Alcohol Consumption' says: 'Results indicated that personal values can serve as significant predictors of the attitudes college students have toward alcohol use, which in turn can predict intentions to drink. Results also indicated that intentions to drink are strongly related to actual alcohol consumption'.

Dougherty, Donald M., James M. Bjork, Robert H. Bennett and F. Gerard Moeller (1999). 'The Effects of a Cumulative Alcohol Dosing Procedure on Laboratory Aggression in Women and Men'. *Journal of Studies on Alcohol and Drugs* 60 (3): 322–9.

Giancola, Peter R. (2002). 'Irritability, Acute Alcohol Consumption and Aggressive Behavior in Men and Women'. *Drug and Alcohol Dependence* 68 (3): 263–74.

Parrott, Dominic J., Amos Zeichner, and Dana Stephens (2003). 'Effects of Alcohol, Personality, and Provocation on the Expression of Anger in Men: A Facial Coding Analysis'. *Alcoholism: Clinical and Experimental Research* 27 (6): 937–45.

Streufert, Siegfried, Rosanne Pogash, Daniela Braig, Dennis Gingrich, Anne Kantner, Richard Landis, Lisa Lonardi, John Roache and Walter Severs (1995). 'Alcohol Hangover and Managerial Effectiveness'. *Alcoholism: Clinical and Experimental Research* 19 (5): 1141–6.

Thomas, Suzanne E., Carrie L. Randall, and Maureen H. Carrigan (2003). 'Drinking to Cope in Socially Anxious Individuals: A Controlled Study'. *Alcoholism: Clinical and Experimental Research* 27 (12): 1937–43.

Shim, Soyeon, and Jennifer Maggs (2005). 'A Cognitive and Behavioral Hierarchical Decision-making Model of College Students, Alcohol Consumption'. *Psychology and Marketing* 22 (8): 649–68.

RESEARCH SPOTLIGHT

'AN ODE TO SUBSTANCE USE(R) INTERVENTION FAILURE(S): SUIF'

by Shlomo Stan Einstein (published in *Substance Use and Misuse*, 2012)

THE VEGETARIAN WHO ATE A SAUSAGE WITH CURRY SAUCE

A report called 'The Vegetarian Who Ate a Sausage with Curry Sauce' can provide cheer both for meat-eaters – because it tells how a hunk of processed meat served as a helpful warning beacon, possibly lengthening a person's life – and for vegetarians – because that meat was the stark symbol of someone's health going very wrong.

Despite its children's-bookish title, this report was published in the January 2003 issue of *The Lancet Neurology*, along with the less child-friendly sounding 'Principles of Frontal Lobe Function' and 'Should I Medicate My Child?'.

The authors, Josef Heckmann, Christoph Lang, Hermann Stefan and Bernhard Neundörfer, of the department of neurology at the University of Erlangen-Nuremberg in Germany, tell the story of a sixty-one-year-old woman who visited her work canteen for lunch with a male colleague. 'The colleague was greatly astonished to see the woman, a passionate vegetarian, order a sausage with curry sauce. Furthermore, during conversation, he noticed that she was a little confused, although she was able to walk and carry her tray without difficulty. Because of the woman's changed behaviour, her colleague arranged a transfer to a hospital for her.'

Heckmann, Lang, Stefan and Neundörfer ran tests, and diagnosed this as a previously unsuspected case of non-convulsive status epilepticus, a form of epilepsy that can lead to coma and other bad states. They gave her medicine to prevent further seizures.

When a sausage induces a visit to hospital, most commonly it is food poisoning that drives the action. But occasionally, the cause and effect are mechanical. A sausage can get stuck on the journey from mouth to stomach, and sometimes does.

Such a case was reported in 2001 at James Paget hospital in Great Yarmouth. As reported by O.N. Enwo and M. Wright in the *International Journal of Clinical Practice*, under the headline 'Sausage Asphyxia': 'a case of supraglottic impaction of the larynx by a piece of sausage occurred in our hospital; the patient was semiconscious. It was managed successfully by a carefully timed laryngoscope blade being inserted into the mouth without the aid of sedative drugs'.

There is some tendency for these cases to occur, or at least to be written up when they occur, in Germany and Austria. One fairly typical report was published in the medical journal *Medizinische Klinik* in 2009 about a seventeen-year-old fellow in Regensburg, Germany, who had a Wiener sausage in his oesophagus.

Occasionally, the mechanical problem occurs elsewhere than in the digestive tract. Austria supplied a new, perfect amplifier for the old saying 'you don't want to know how the sausages are made'. The June 2006 issue of the journal *Wiener Klinische Wochenschrift* [the *Central European Journal of Medicine*] featured a photo-illustrated report, from the city of Linz, called 'Finger Amputation by a Sausage Packing Machine'.

Heckmann, Josef, Christoph Lang, Hermann Stefan, and Bernhard Neundörfer (2003). 'The Vegetarian Who Ate a Sausage with Curry Sauce'. *Lancet Neurology* 2 (1): 62.

Enwo, O.N., and M. Wright (2001). 'Sausage Asphyxia'. *International Journal of Clinical Practice* 55 (10): 723–4.

Gäbele, Erwin, Esther Endlicher, Ina Zuber-Jerger, Wibke Uller, Fabian Eder and Jürgen Schölmerich (2009). 'Impaction of a "Sausage Bread" in the Oesophagus: First Manifestation of an Eosinophilic Oesophagitis in a 17-Year-Old Patient'. *Medizinische Klinik* 104 (5): 386–91.

Huemer, Georg M., Harald Schoffl and Karin M. Dunst (2006). 'Finger Amputation by a Sausage Packing Machine'. *Wiener Klinische Wochenschrift* 118 (11/12): 321.

MAY WE RECOMMEND

'HOT POTATOES IN THE GRAY LITERATURE'
by Brian Pon and Alan Meier (published as *Recent Research in the Building Energy Analysis Group*, 1993)

MINCING VEGETARIANISM, NOT WORDS

Vegetarianism – the wanton ingestion of nothing but non-meat – sometimes produces or provokes antipathy, hostility and disgust. Researchers have struggled to understand why. In 1945, as the Second World War was ending, US Army Major Hyman S. Barahal, chief of the psychiatry section of Mason General Hospital in Brentwood, New York, issued a report called 'The Cruel Vegetarian'. Major Barahal began by explaining the word 'vegetarianism' for anyone who might be ignorant or confused: 'It consists essentially in the exclusion of flesh, fowl and fish from the dietary.'

Major Barahal drew upon his own experience at having met, and endured the presence of, several vegetarians. 'Their exaggerated concern over the welfare of animals betrays the utter contempt and hatred which they hold for the human race generally', he reported. 'As far as the present writer knows, no [previous] article has ever attempted to explain the psychology of a person who, of his own free will, becomes a fervent follower of the cult.'

THE CRUEL VEGETARIAN*

BY MAJOR HYMAN S. BARAHAL, M. C.†

In recent years there has been a growing popular interest in psychology and human behavior. Various phases of normal and abnormal psychology are taught in our schools, discussed in our homes, and utilized in a practical way in better adjusting ourselves to our environment. Many business establishments have instituted psychiatric departments to aid them in the hiring of employees as well as in handling emotional problems which may arise.

Major Barahal preferred to mince vegetarians, rather then words. He cut directly to the meat of the matter: 'The average vegetarian is eccentric, not only as regards his food, but in many other spheres as well. Careful observation of his views ... will frequently reveal somewhat twisted and rather peculiar attitudes and prejudices. In short, the average vegetarian is not definitely "a lunatic", but he certainly fringes on it.' His report carries a footnote that says little but implies much: 'This manuscript was prepared prior to the author's entering military service and contains no material or conclusions gained from experiences in military service.'

Sixty years later, in 2005, Daniel Fessler of the University of California, Los Angeles, and three colleagues looked at the emotional tangle provoked in vegetarians by vegetarianism's opposite number, meat eating.

Their treatise, called 'Disgust Sensitivity and Meat Consumption: A Test of an Emotivist Account of Moral Vegetarianism', appeared in the journal *Appetite*. It contrasts 'moral vegetarians', who 'view meat avoidance as a moral imperative', with 'health vegetarians', who 'are upset by others' meat consumption'. Fessler and his colleagues reached an emphatic conclusion: 'moral vegetarians' disgust reactions to meat are caused by, rather than causal of, their moral beliefs'.

The same year, 2005, the *British Journal of Nutrition* served up a French study called 'Emotions Generated by Meat and Other Food Products in Women', which includes a fairly rare occurrence of the phrase 'low meat-eating women': 'As expected, the low meat-eating women felt more disappointment, indifference and less satisfaction towards meat than did the high meat-eating women. However, the low meat-eating women also stated other negative emotions such as doubt towards some starchy foods. The only foods that they liked more than high meat-eating women were pears and French beans.'

Researchers at the Institut National de la Recherche Agronomique, the Centre Européen des Sciences du Goût and the Laboratoire de Psychologie Sociale et Cognitive collaborated in (1) collecting photographs of pears, apples, rice, pasta, pizza, pound cake, turkey, rabbit, pork chops, offal and other kinds of food; and (2) gathering sixty French women; and then (3) exposing the latter to the former. The women's reactions to those photos led the researchers to a conclusion that perhaps did not surprise them: the failure to eat lots of meat is 'associated with specific negative emotions regarding meat and other foods'.

Barahal, Hyman S. (1946). 'The Cruel Vegetarian'. *Psychiatric Quarterly* 20 (1 supplement): 3–13.

Fessler, Daniel M.T., Alexander P. Arguello, Jeannette M. Mekdara and Ramon Macias (2003). 'Disgust Sensitivity and Meat Consumption: A Test of an Emotivist Account of Moral Vegetarianism'. *Appetite* 41 (1): 31–41.

Rousset, S., V. Deiss, E. Juillard, P. Schlich and S. Droit-Volet (2005). 'Emotions Generated by Meat and Other Food Products in Women'. *British Journal of Nutrition* 94 (4): 609–19.

CALL FOR OPINIONS

In the years since Major Barahal's report on the psychology of vegetarians, questions have been raised about the definition of 'vegetarianism' used as a basis of his incisive analysis. One question in particular has caused agony for the *Annals of Improbable Research* scientific survey team.

The Question: Typically, strict vegetarians avoid eating meat, but consider anything else to be fair game (so to speak). Biologists classify most bacteria as being neither animal nor vegetable. If offered a good home-cooked meal of baked, stuffed bacteria, what's a strict vegetarian to do? Are strict vegetarians allowed to eat bacteria?

The Results: An initial survey, conducted in 1999, chewed on the matter. Respondents' answers ranged from 'In a 5-kingdom classification system, vegetarians are allowed to consume 80% of living things' to 'Their GI tracts should be purged of their microfauna... Then see how healthy they are'. In summary, 32 percent of respondents allowed bacteria consumption and 32 percent did not, with 36 percent suggesting 'other' allowances.

Conclusions: Does this remain the state of this vego-bacterial matter? The *Annals of Improbable Research* team have been inspired by the January/February 2013 edition of the journal *Gastrointestinal Cancer Research* in which the editor-in-chief Daniel G. Haller issued 'A Call for Opinions'. Haller opines: 'Opinions from experts – however designated – remain important' and 'The worst ending to any editorial is "Further work needs to be done".' Further work needs to be done.

To propose your answer to the question, 'Are strict vegetarians allowed to eat bacteria?', please send by email:

- a photograph that clearly provides evidence of (1) your standing as a vegetarian, carnivore or omnivore; (2) a good home-cooked meal, noting whether it was digested with or without bacteria
- a current curriculum vita outlining your credentials as a scientist
- a pithy statement hashing out your opinion and rehashing your expert qualifications, both culinary and scholarly.

Send to marca@improbable.com with the subject line:
BACTERIA DIGEST SURVEY

SCRUTINIZED FALAFEL

A study called 'Effect of a Popular Middle Eastern Food (Falafel) on Rat Liver' is available to the public. Focusing strictly on the medical consequences (for rats) of eating 'chickpeas paste seasoned with garlic, parsley, and special spices, then deep fried in vegetable oil', it's a ten-page journey from delight to despair, and finally to indifference.

Sana Janakat and Mohammad Al-Khateeb of the Jordan University of Science and Technology in Irbid, Jordan, wrote the report. It somewhat enlivens the February 2011 issue of the journal *Toxicological & Environmental Chemistry*.

Janakat and Al-Khateeb start with some cheery praise: 'Falafel is considered the most popular fried food by all socio-economic classes in most Middle Eastern countries. It is consumed for breakfast, dinner, or as a snack. Among low-income families and labourers it is consumed on a daily basis, due to its availability, relatively low price, and good taste.'

Then come several pages of unhappy news that lead to a depressing and technical conclusion: 'Long-term consumption of falafel patties caused a significant increase in ALP [alkaline phosphatase], ALT [alanine transaminase], bilirubin level and increased liver weight/body weight ratio ... This indicates that consumption of large amounts of falafel on daily basis might lead to hepatotoxicity.'

But wait! That's not the end of the story.

Here's what Janakat and Al-Khateeb say they did.

First they gathered falafel: 'Frying oil samples and falafel patties were collected from 20 restaurants located in different socio-economic neighbourhoods from the city of Irbid.' They homogenized the falafel, soaked it for twenty-four hours, filtered it through cheesecloth and centrifuged it. This produced the experimental material – concentrated falafel – with which they performed two experiments.

Sample collection

Frying oil samples and falafel patties were collected from 20 restaurants located in different socio-economic neighborhoods from the city of Irbid with the help of a health inspection team of the municipality of Irbid. Oil discoloration was assessed visually, and the quality of falafel frying oil and of oil extracted from falafel patties was assessed using lipid peroxidation (LPO) induction in rat liver homogenate prepared using $FeCl_2$/ Ascorbate (Ohkawa, Ohishi, and Yagi 1979). For long-term experiments, falafel patties were bought from one of the restaurants with an average work load, medium oil discoloration and medium LPO, prepared in frying oil and 3 days old; frying oil was used for 5 days in most restaurants.

An explanation of how and which falafel was collected

The first experiment assessed the short-term effect of eating falafel, short term in this case meaning five days. Janakat and Al-Khateeb extracted the oil from some falafel patties, and force-fed it to some rats for the whole five days. Then they killed those rats, and did post-mortems to get at the livers. The livers (happily, in a sense) looked in pretty good shape. Thus, the report says, the short-term effects of eating falafel are pretty benign.

The second experiment aimed to clarify the long-term effect of falafel consumption. A fresh batch of rats got to eat lots of falafel – as much as they pleased, whenever they wanted it – for a month. That was their entire diet: falafel, falafel, falafel. Then they were killed. Here, the post-mortem results were ugly. The study intones that 'long-term consumption of falafel patties (30 days) caused yellowish discoloration of the liver distinctive of liver necrosis', suggesting that 'the consumption of falafel as the sole source of nutrition for a long period of time … can generate a hepatotoxic effect leading to liver necrosis'.

That may sound like bad news, but apparently it's not. The very next sentence – nearly the last thing said in the report – is this: 'Falafel consumption in moderation and in conjunction with other food items or beverages containing high antioxidant levels can be considered as safe.'

Janakat, Sana M., and Mohammad A. Al-Khateeb (2011). 'Effect of a Popular Middle Eastern Food (Falafel) on Rat Liver'. *Toxicological & Environmental Chemistry* 93 (2): 360–9.

WATER TASTES LIKE WATER

Can people perceive the difference between bottled water and tap water? Two studies suggest that – at least in France and in Northern Ireland – water tastes like water.

A French research team ran a taste test of six different bottled mineral waters and six municipal tap waters. This was a collaboration between two scientific institutes (CNRS, the French National Centre for Scientific Research, and INRA, the National Institute of Agronomic Research) and Lyonnaise des Eaux, a company that manages many public water supplies.

The tasters were '389 persons from all over France'. The bottled waters were 'chosen among the French bottled water available'. The tap waters were 'from various regions of France, supplied by Lyonnaise des Eaux'. The team published their study in the *Journal of Sensory Studies* in 2010.

The researchers identified what they call the 'three main tastes of water' that can be found if one swigs a great variety of bottled and tap waters. These are 'the bitterness of poor mineralised water, the neutral taste (associated with coolness) of water with medium mineralisation and the saltiness and astringency of highly mineralised water.' The report concludes that 'most consumers cannot distinguish between bottled water and tap water when the latter is chlorine-free'. But most is not all. The report goes on to say: 'However, 36% of the subjects were found able to distinguish between tap water and bottled water.'

Five years earlier, in Belfast, Deborah Wells had run a similar test, with slightly more than one thousand people tasting water from several sources. These included 'one of the UK's most popular brands of still bottled mineral water (Evian, Danone Waters), distilled water (supplied by Queen's University Belfast), and tap water (supplied to Belfast by the Water Service from the

Silent Valley, Co. Down)'. Wells, a senior lecturer in psychology at Queen's University, then wrote a report called 'The Identification and Perception of Bottled Water', which appeared in the journal *Perception*.

'The findings from this study indicate that people cannot correctly identify bottled water on the basis of its flavour', she declares. This 'suggests that the currently high consumer demand for this beverage must be based on factors other than taste or olfactory perception'.

That thought was not entirely new.

In 2004, we awarded the Ig Nobel Prize in chemistry to the Coca-Cola Company of Great Britain. The company has long publicized the existence of a 'secret formula' for its signature cola beverage. The key ingredient is water. But that's not what won them the prize. They were honoured for using advanced technology (mostly pumps) to convert ordinary tap water (obtained from Thames Water, in Sidcup) into Dasani, a pricey, water-filled bottle. It sold briskly – until news broke that Dasani was just bottled tap water.

But … it wasn't 'just' bottled tap water. As the *Guardian* reported on 20 March of that year: 'The entire UK supply of Dasani was pulled off the shelves because it has been contaminated with bromate, a cancer-causing chemical.'

Teillet, Eric, Christine Urbano, Sylvie Cordelle and Pascal Schlich (2010). 'Consumer Perception and Preference of Bottled and Tap Water'. *Journal of Sensory Studies* 25 (3): 463–80.
Wells, Deborah L. (2005). 'The Identification and Perception of Bottled Water'. *Perception* 34 (10): 1291–2.

IN BRIEF

'METAL DETECTORS: AN ALTERNATIVE APPROACH TO THE EVALUATION OF COIN INGESTIONS IN CHILDREN?'
by S.P. Ros and F. Cetta (published in *Pediatric Emergency Care*, 1992)

BIG BELLY DATA

> ## Taking the biscuit: the structure of British meals
> Mary Douglas and Michael Nicod
>
> The way in which the British eat is as formally structured as a Bach sonata. But however the composition begins, there's one coda: the biscuit.

Confident that no one would notice what he was doing, Michael Nicod spent months in the homes of families he did not know, making detailed notes about everything they ate. Nicod was performing research for Britain's Department of Health and Social Security in 1974. He and his colleague, University College London professor Mary Douglas, wrote a report called 'Taking the Biscuit: The Structure of British Meals'.

Nicod and Douglas wanted to identify what typical British persons see as the essential parts of their typical meals. The pair drew on their training as anthropologists: 'We imagined a dietician in an unknown Papuan or African tribe wondering how to introduce a new, reinforcing element into tribal diet. We assumed that the dietician's first task would be to discover how the tribe "structured" their food.'

Second Correspondence
Course two repeats structure of course one in different materials

Mode	Structure	Elements
Course 1		
hot and savoury	staple	potato
	centre	meat, fish, egg
	trimming	green veg, stuffing, Yorkshire pudding
	dressing	thick brown gravy
Course 2		
hot or cold, sweet	staple	cereal
	centre	liquid custard
	dressing	liquid custard or cream

Fourth Correspondence

Meal B repeats meal A in course sequence, but keeps to the staple of course two

Mode	Structure	Elements
Course 1		
savoury, hot or cold	staple	bread
	centre	meat, fish or egg or baked beans
	trimmings	optional
Course 2		
sweet, cold	staple	bread
	centre	jam
	trimmings	butter
Course 3		
hot sweet drink	optional cake for Sundays or biscuits	

Table 3 versus Table 5 from 'Taking the Biscuit'. Say the authors: 'It is no surprise to the native Englishman that the distinction between hot and cold is much valued in this dietary system.'

Nicod lived as a lodger with 'four working-class families where the head was engaged in unskilled manual labour', in East Finchley, Durham, Birmingham and Coventry. He stayed in each place at least a month, 'watching every mouthful and sharing whenever possible'.

Nicod and Douglas express confidence in the obliviousness of the natives. 'We reckon', they write, 'that after 10 days of such a discreet and incurious presence, the most sensitive housewife, busy with her children, settles down to her routine menus.'

Many others have imagined new ways to examine and change the eating habits of persons other than themselves.

In 2008, Mariana Simons-Nikolova and Maarten Bodlaender of the Netherlands applied for a patent for an electro-mechanical process they call 'Modifying a Person's Eating and Activity Habits'. Their video/computer system would monitor an individual's head and hands to detect when they were eating. It would then announce to them via the TV or computer, 'You

Are Now Eating'. Simons-Nikolova and Bodlaender explain: 'By providing the feedback when the subject is still eating or drinking, the subject is helped to stop the eating or drinking sooner than if no feedback had been given.'

A 2008 patent by three Israeli inventors describes 'a sensor which detects: (a) the patient swallowing, (b) the filling of the patient's stomach, and/or (c) the onset of contractions in the stomach as a result of eating'. Electric current can then, for dietary reasons, be 'driven into muscle tissue of the subject's stomach'. This 'induces in the subject a sensation of satiation, discomfort, nausea, or vertigo'. The entire plan, they say, pertains to 'appetite regulation, and specifically to invasive techniques and apparatus for appetite control and treating obesity'.

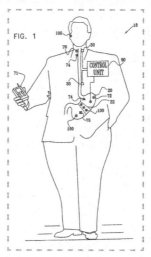

Fig. 1 of 15 from the 2008 patent 'Regulation of Eating Habits'

Three Indian inventors filed a patent application in 2010 for a 'refrigerator for obese persons'. The fridge monitors 'all eating and drinking', and dispenses diet advice. Also, 'a reflecting mirror film on the door makes the person to control overeating as soon as he stands before the fridge'.

With these and related plans society becomes more equipped for 'watching every mouthful' of some of its members.

Douglas, Mary, and Michael Nicod (1974). 'Taking the Biscuit: The Structure of British Meals'. *New Society* 19: 744–77.

Simons-Nikolova, Mariana, and Maarten P. Bodlaender (2008). 'Modifying a Person's Eating and Activity Habits'. European patent no. 1,964,011, 3 September.

Aviv, Ricardo, Ophir Bitton and Shai Policker (2008). 'Regulation of Eating Habits'. US patent no. 7,437,195, 14 October.

Anandampillai, Aparna Thirumalai, Vijayan Thirumalai Anandampillai and Anandvishnu Thirumalai Anandampillai (2010). 'Refrigerator for Obese Persons'. US patent application 12/799,645, 14 October.

MAY WE RECOMMEND

'EXPONENTIAL NONNEGATIVITY ON THE ICE CREAM CONE'
by Ronald J. Stern and Henry Wolkowicz (published in *SIAM: Journal on Matrix Analysis and Applications*, 1991)

GRUEL WORLD

Our relationship with cooked cereal owes much to Louis J. Lee of Rochester, New York. Thanks to him, we no longer need to chew the stuff as much as we once had to.

Lee solved a problem that he described in 1963 in a US patent application. He explained that cooked cereals 'tend to become pasty on cooking and to lose particle texture and flavor on prolonged heating ... In many commercial eating establishments, particularly in cafeterias, it is customary to cook up a large batch of a cooked cereal ... After several hours on a steam table it is not unusual for a batch of cooked cereal to become a congealed, gelatinous mass. As a result, the batch is unappetizing in appearance and taste and usually is dumped into a garbage can without further ado.'

Lee also gave us a handy, professional definition of 'cooked cereals'. These, he wrote, 'are prepared foodstuffs of grain ... which usually must be cooked for a short period of time in boiling water in order to be in a normally edible condition ...

cooking causes the dehydrated grain particles to absorb water and to soften, making the same palatable and digestible.'

Mr Lee is an unsung hero of modern quick-cooking hot cereal. His innovation was, from a chemist's point of view, simple (though some breakfasters may find it rather over-syllabic for their taste). The Lee method is to cook the cereal mixed together 'with an edible monoglyceride of the chemically saturated type'.

Three decades earlier, William C. Baxter of Newtown, Connecticut, looked at a different aspect of the cereal-chewiness problem. Baxter found out what happens when you combine shredded wheat with ice cream.

This hybrid foodstuff, Baxter pointed out in a patent obtained in 1936, 'is really delectable only when the cereal pieces, when taken into the mouth with an intermingled mass of ice cream, are in fairly crisp condition, by which I mean a condition such that these cereal pieces are not yet soggy, although somewhat moist along and immediately below their superficies'.

Baxter's patent might, in the long run, be best remembered for one poetical passage. Read it aloud to your family each day over breakfast (or, even better, confide it to some stranger at your coffee shop): 'In the ordinary eating of shredded wheat biscuit with milk or cream, if a thin milk is used and the biscuit is allowed to soak even for a comparatively short time in a pool of such milk, the disk is an unappetizing and highly unsatisfactory one, except to those who for lack of teeth or otherwise enjoy mush, whereas, if the lacteal fluid employed is a heavy or medium cream and if the biscuit and such fluid are together consumed even by a very slow eater, the pieces of the biscuits spooned off or otherwise segregated for each bite carry suitable portions of the lacteal fluid on their surfaces and within their interstices and yet there is a certain definite and desirable crisp-chewability retained by said pieces.'

Lee, Louis J. (1963). 'Process for Preparing a Cooked Cereal and the Resulting Product'. US patent no. 3,113,868, 10 December.

Baxter, William C. (1936). 'Food Product and Method of Making Same'. US patent no. 2,065,550, 29 December.

FURTHER CEREAL STUDIES: MILK, THEN WATER, THEN PLIERS, THEN MILK

After generations of humans had been pouring cows' milk onto breakfast cereal flakes and then pouring that milk/flake mixture into themselves, a researcher named Luigi Degano fed breakfast cereal to twenty-one cows in Italy. Degano wanted to see how this might affect the milk that later issued from the cows.

Degano, based at the Istituto Sperimentale Lattiero Caseario in Milan, published the results of this feed-flakes-to-cows experiment in 1993 in the journal *Tecnica Molitoria*.

Degano called the study 'Cereal Flakes in Milk Cows Diet: Effects on Yield and Milk Quality'. He reported that yes, mixing plenty of maize-and-barley flakes into the cows' usual, unflaked maize-and-barley fodder did result in different milk. Slightly different. Those cows gave about 2 percent more milk (by volume), with about 2 percent richer protein content and about 2 percent greater creaminess. All this as compared with the milk-making of twenty-one cows that munched only the usual mealy mush.

Degano's monograph seems to have attracted little attention, at least in print, from other dairy scientists. And it garnered just about no acclaim from the general public in Italy or abroad.

Scientists have, as a group, shown more interest in cereal's crispness, especially as it interacts with liquid, than in how the flakes interact with cows or with human innards.

The most famous report, 'A Study of the Effects of Water Content on the Compaction Behaviour of Breakfast Cereal Flakes', was published in 1994, in the journal *Powder Technology*. Three scientists at the Institute of Food Research in Norwich wrote it.

In 2001, a student named Kunchalee Luechapattanapom, at the Asian Institute of Technology in Bangkok, Thailand, submitted a master's thesis entitled 'Acoustic Testing for Evaluating the Crispness of Breakfast Cereals'. Luechapattanapom describes the basics, in this physico-mathematical passage: 'Three brands of breakfast cereals, i.e. Kellogg's corn flake, Honey Stars and Koko Krunch were evaluated their crispness using acoustic testing and mechanical testing. The sound was produced by crushing the samples with the spring loaded pliers. The original amplitude-time curves were converted to the power spectrum of frequencies by using Fast Fourier Transform (FFT).'

Others, too, tried their hand at the flake analysis game, generally using either the Norwich or the Bangkok approach.

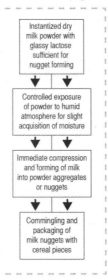

Steps for forming novel dry milk pieces

But Lawrence Edward Bodkin Sr, an inventor in Jacksonville, Florida, may have circumvented such traditional worries about milk and breakfast cereal flakes, by combining the two elements into one. In 1998, Bodkin patented a foodstuff he calls 'Breakfast Cereal with Milk Pieces'. Bodkin's odd patent describes a 'commingling and packaging of milk nuggets with cereal pieces … The milk pieces may be compact, or flattened and flake shaped and may generally be as variable as the shapes of the cereal'. Any strangeness in the milk's flavour, he writes, 'is unlikely to be noticed due to the typically more dominant flavors of the cereal'.

Drawing: 'a quantity of prepared cereal, in a popular shape containing milk aggregates or milk pieces of a size and shape comparable to those of the cereal'

For their rigorous analysis of soggy breakfast cereal, Georget, Parker and Smith were awarded the 1995 Ig Nobel Prize in physics.

Degano, Luigi (1993). 'Cereal Flakes in Milk Cows Diet: Effects on Yield and Milk Quality'. *Tecnica Molitoria* 44 (1): 5–14.

Georget, Dominique M.R., Roger. Parker, and Andrew C. Smith (1994). 'A Study of the Effects of Water Content on the Compaction Behaviour of Breakfast Cereal Flakes'. *Powder Technology* 81 (2): 189–95.

Luechapattanapom, Kunchalee (2001). 'Acoustic Testing for Evaluating the Crispness of Breakfast Cereals'. Thesis submitted in the partial fulfilment of the requirements for the degree of master of science, Asian Institute of Technology, School of Environment, Resources and Development, Bangkok, Thailand, August.

Bodkin Sr, Lawrence Edward (1998). 'Breakfast Cereal with Milk Pieces'. US patent no. 5,827,564, 27 October.

METICULOUSLY SPIRITED HANDWRITING

A research project in Turkey examined whether and how drinking affects the quality of one's writing. This was 'hard' rather than 'soft' science – it ignored anything fuzzy and hard-to-measure, such as the literary quality of the writing, or its emotional content. The experiment focused, with great discipline, on something that can be gauged more objectively: the extent to which drinking makes people's penmanship go wobbly.

The goal: to establish that a sometimes-suspect criminal justice tool is dependable, accurate and precise.

Faruk Acolu, at the Council of Forensic Medicine, in Istanbul, and Nurten Turan, at the University of Istanbul, published their study in the journal *Forensic Science International*. It is called 'Handwriting Changes Under the Effect of Alcohol'.

They describe going to the annual party of one of Turkey's most prominent companies, soliciting volunteers from among the attendees. Each volunteer took a breath test and then filled out a questionnaire. 'Two of the participants', the report reveals, 'could not complete the text after consumption of alcohol, and therefore they were excluded.'

Handwriting, when sober (top) and under the influence (bottom)

Seventy-three volunteers did go through the full rigours of the experiment. Each sat at a well-lit desk and copied out a standard passage of text onto an unlined pad of paper. They did it when sober, and then again when intoxicated.

To achieve intoxication, the report explains, 'The participants consumed ethyl alcohol without limitation'. Each individual was permitted to select and guzzle his or her favourite kind of drink. Twenty-three of them attained a condition of moderate tipsiness (with alcohol-in-their-blood levels of less than fifty milligrams per one hundred millilitres – 50 mg/100 ml). Twenty-four became middling tipsy (with levels between 51mg/100ml and 100mg/100ml). The other twenty-six got sloshed.

Acolu and Turan then compared the writing samples done while sober with those produced under the influence of drink. They used good equipment – an Olympus X-Tr stereo micro-scope, with direct and oblique angle lighting, and a VSC 2000 Foster and Freeman video spectral comparator. Each handwriting sample yielded up a twenty-six-item checklist full of data: word length, height of upper-case character bodies, variation in spacing between words, number of visible 'tremors', number of misspellings and so on. The published report includes, with the statistics, some lovely photos of before-and-after samples.

Acolu and Turan found pretty much what they had hoped to find – evidence that most people's handwriting gets worse and worse as they become more and more intoxicated. However, they write – with perhaps just a hint of disapproval or maybe even disappointment – that 'the writing quality of a few seemed to get better after alcohol consumption'.

Range of Changing Handwriting Parameters under the Influence of Alcohol

Variables	Unchanged (%)	Decreased (%)	Increased (%)
Word length	1.07	20.36	78.57
Height of upper case	1.37	27.40	71.23
Height of lower case	1.37	31.51	67.12
Height of ascending letter	1.37	32.88	65.75
Height of descending letter	0.00	28.77	71.23
Spacing between words	0.00	24.70	75.30
Number of tremor	17.81	20.55	61.64
Number of angularity	32.88	21.92	45.20
Number of tapered ends	1.37	23.29	75.34

Drunk handwriting, by the numbers

Why go to all this trouble? Because, Acolu and Turan say, the results of previous studies done by other people 'are mostly not based upon statistical data and [are] therefore unsatisfying'.

Does the report constitute proof that handwriting analysts can reliably discern the relationship between penmanship and drunkenness? No. But for anyone who takes drunken handwriting seriously, it's a step in the right direction.

Acolu, Faruk, and Nurten Turan (2003). 'Handwriting Changes Under the Effect of Alcohol'. *Forensic Science International* 132 (3): 201–10.

SCIENTISTS DOWN THE PUB

To answer the question, 'What happens when people drink alcohol?' one can read through thousands of research studies published in respected scholarly journals. One must look a bit harder to answer a different question: 'What, exactly, did some of those researchers hope to learn by doing that research?'

Let's take a quick hop through the literature in one publication – the *Journal of Studies on Alcohol and Drugs*, which boasts of being 'the oldest alcohol/addiction research journal currently published in the United States'. It started life in 1940 as the *Quarterly Journal of Studies on Alcohol*, then adjusted and re-adjusted its name as research funders changed their focus or preferred vocabulary.

A study called 'Observational Study of Alcohol Consumption in Natural Settings: The Vancouver Beer Parlor' appeared in a 1975 issue. Authors Ronald Cutler and Thomas Storm, at the University of British Columbia, say they visited 'approximately 25' Vancouver beer parlours, wherein they observed the patrons. They distil what they learned into three thoughts: 1) people drank at a 'relatively constant' rate; 2) the longer people spent drinking, the more they drank; and 3) in bigger groups, people spent more time drinking, and so drank more drinks. Cutler and Storm explain that 'these findings are consistent with' those reported ten years earlier in a study called 'The Isolated Drinker in the Edmonton Beer Parlor'. Cutler and Storm say the Edmonton researcher behind that report, R. Sommer, 'found that patrons drinking alone ordered an average of 1.7 glasses of beer, while patrons drinking with a group ordered an average of 3.5 glasses. This difference was accounted for by the length of time solitary and group drinkers spent in the beer parlour and not by the rate of drinking.'

Cutler and Storm performed additional research. Perhaps their most ambitious work, called 'Observations of Drinking in Natural Settings: Vancouver Beer Parlors and Cocktail Lounges', published in 1981, says: 'The question that, to a great extent, motivated this study was: how much do people actually drink in drinking establishments which are most heavily patronised by ordinary social drinkers? The answer is, in qualitative terms: a fair number of people drink quite a lot, especially in beer parlours, and particularly when the group is large.'

Jump ahead to July 2012 and one finds a study, entitled 'Daily Variations in Spring Break Alcohol and Sexual Behaviors Based on Intentions, Perceived Norms, and Daily Trip Context', by Megan Patrick, of the University of Michigan, and Christine Lee, of the University of Washington in Seattle. Patrick and Lee gathered information from 261 American university students, from which they learned this: 'Students who went on longer trips, who previously engaged in more heavy episodic drinking, or who had greater pre–Spring Break intentions to drink reported greater alcohol use during Spring Break. Similarly, students with greater pre–Spring Break intentions to have sex, greater perceived norms for sex, or more previous sexual partners had greater odds of having sex.'

Cutler, Ronald E., and Thomas Storm (1975). 'Observational Study of Alcohol Consumption in Natural Settings: The Vancouver Beer Parlor'. *Journal of Studies on Alcohol and Drugs* 36 (9): 1173–83.

Sommer, R. (1965). 'The Isolated Drinker in the Edmonton Beer Parlor'. *Quarterly Journal of Studies on Alcohol* 26 (1): 95–110.

Storm, Thomas, and Ronald E. Cutler (1981). 'Observations of Drinking in Natural Settings: Vancouver Beer Parlors and Cocktail Lounges'. *Journal of Studies on Alcohol and Drugs* 42 (11): 972–97.
Patrick, Megan E., and Christine M. Lee (2012). 'Daily Variations in Spring Break Alcohol and Sexual Behaviors Based on Intentions, Perceived Norms, and Daily Trip Context'. *Journal of Studies on Alcohol and Drugs* 73 (4): 591–6.

MAY WE RECOMMEND

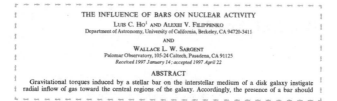

'THE INFLUENCE OF BARS ON NUCLEAR ACTIVITY'
by Luis C. Ho, Alexei V. Filippenko and Wallace L.W. Sargent
(published in the *Astrophysical Journal*, 1997)

The authors, at the University of California, Berkeley, and Palomar Observatory, report that: 'The presence of a bar seems to have no noticeable impact on the likelihood of a galaxy to host either nuclear star formation or an active galactic nuclei.'

SEVEN

SOME THINGS STICK OUT

MAY WE RECOMMEND

Necessary and Sufficient Conditions for Bang–Bang Control

H. FUJIHIRA[1]

Communicated by L. Cesari

Abstract. Necessary and sufficient conditions for the optimal control to be bang–bang are presented for a nonlinear system. The payoff,

'NECESSARY AND SUFFICIENT CONDITIONS FOR BANG-BANG CONTROL'
by H. Fujihira (published in the *Journal of Optimization Theory and Applications*, 1978)

Some of what's in this chapter: Half a penis, half a penis, onwards (duck permitting) • Fingers, sex and ultimatums • Historic sex video: The inside story • Her dirty image problem • Outcomes of eye comes • When his length was excessive • Aggressive nipple assumptions • The British/Argentinian protuberance disagreement • Long words and naughty words • Budongoan chimp johnson-cleaning • Questioning the Menstrual Joy Questionnaire • Sexual conquests of artists • Reproduction of reproduction parts • Immoral thoughts on the side • Informatics for official fake prostitutes

UNDER THE KNIFE

About once per decade, the medical profession takes a careful look back at Thailand's plethora of penile amputations. The first great reckoning appeared in a 1983 issue of the *American Journal of Surgery*. 'Surgical Management of an Epidemic of Penile Amputations in Siam', by Kasian Bhanganada and six fellow physicians at Siriraj Hospital in Bangkok, introduces the subject: 'It became fashionable in the decade after 1970 for the humiliated Thai wife to wait until her [philandering] husband fell asleep so that she could quickly sever his penis with a kitchen knife. A traditional Thai home is elevated on pilings and the windows are open to allow for ventilation. The area under the house is the home of the family pigs, chickens, and ducks. Thus, it is quite usual that an amputated penis is tossed out of an open window, where it may be captured by a duck.'

The report explains, for readers in other countries: 'The Thai saying, "I better get home or the ducks will have something to eat", is therefore a common joke and immediately understood at all levels of society.'

The bulk of the paper reports how the doctors and their colleagues learned, over the course of attempting eighteen reimplantations, how to improve the necessary surgical techniques. Unambiguous photographs supplement the text.

'Interestingly', the physicians remark at the very end, 'none of our patients filed a criminal complaint against their attackers.'

An article called 'Factors Associated with Penile Amputation in Thailand', published in 1998 in the journal *Nursing Connections*, explores the reasons behind that. Gregory Bechtel and Cecilia Tiller, from the Medical College of Georgia (in Atlanta), gathered data from three couples who had been part of the epidemic. The couples, by then divorced, discussed their experience calmly. Bechtel and Tiller report that in each case, three things had happened during the week prior to dismemberment: (1) a financial crisis; (2) 'ingestion of drugs or alcohol by the husband immediately prior to the event'; and (3) 'public humiliation of the wife owing to the presence of a second "wife" or concubine'.

In 2008, the *Journal of Urology* carried a retrospective by Drs Genoa Ferguson and Steven Brandes of the Washington University in St. Louis, called 'The Epidemic of Penile Amputation in Thailand in the 1970s'. Ferguson and Brandes conclude that: 'Women publicly encouraging and inciting other scorned women to commit this act worsened the epidemic. The vast majority of worldwide reports of penile replantation, to this day, are a result of what became a trendy form of retribution in a country in which fidelity is a strongly appreciated value.'

In 2013, Bhanganada, Chayavatana, Pongnumkul, Tonmukayakul, Sakolsatayadorn, Komaratat and Wilde were awarded the Ig Nobel Prize in public health, which they graciously accepted.

Bhanganada, Kasian, Tu Chayavatana, Chumporn Pongnumkul, Anunt Tonmukayakul, Piyasakol Sakolsatayadorn, Krit Komaratat and Henry Wilde (1983). 'Surgical Management of an Epidemic of Penile Amputations in Siam'. *American Journal of Surgery* 146 (3): 376–82.

Bechtel, Gregory A., and Cecilia M. Tiller (1998). 'Factors Associated with Penile Amputation in Thailand'. *Nursing Connections* 11 (2): 46–51.

Ferguson, Genoa G., and Steven B. Brandes (2008). 'The Epidemic of Penile Amputation in Thailand in the 1970's'. *Journal of Urology* 179 (4): 312.

RESEARCH SPOTLIGHT

'DIGIT RATIO (2D:4D) MODERATES THE IMPACT OF SEXUAL CUES ON MEN'S DECISIONS IN ULTIMATUM GAMES'
by Bram Van den Bergh and Siegfried Dewitte (published in the *Proceedings of the Royal Society B*, 2006)

THE WILD FRONTIERS IN MEDICINE

Dr Pek van Andel's MRI sex video has thrust its way into an argument that periodically convulses the public and the courts. The video shows the first moving images of a couple's sex organs while those organs were in use. It gives graphic new life to a question as old as sin: what is pornography?

As used by Dr van Andel and his team, the Magnetic Resonance Imaging (MRI) scanner lets us probe anew, and deeply, this legal and philosophical chestnut.

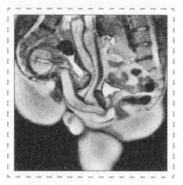

Midsagittal image of anatomy taken during experiment 12

Justice Potter Stewart famously wrote in a 1964 US Supreme Court decision that defining which materials are pornographic is hard, but recognizing them is easy. Quoth the justice: 'I know it when I see it.'

Laypersons watching the van Andel video have a tougher time. During the short time it's been on the Internet, more than 2.5 million people have taken a look. Many, unaccustomed to seeing medical imagery of internal organs, struggled to make sense of the unfamiliar shapes and motions. Their comments about what they saw, posted on YouTube (www.youtube.com/watch?v=OVAdCKaU3vY), make this clear. For every excited *'Agggggggggghhhhhhhhhh!!!!!!!'* there is a baffled *'???'*.

Some people expressed confusion. One wrote: 'Took me a while to figure it out. I thought the man's torso was his penis.'

Another hazarded that: 'The dark spots on either side of the "line" (their skin) are the bladders. The spines are at the outside edges. As best as I can tell it's the womb being bounced around so much.'

A third explained: 'It's obviously missionary. Anyone can see the spines of the man and woman are on the outsides, which shows they are facing each other.'

A number of people do find stimulation, and perhaps even satisfaction, as expressed in this remark: 'It kind of loses something with just the white noise audio ... Having said that, I still need a cigarette now.'

Van Andel made this (possibly) salacious video in the late 1990s, but kept pretty quiet about it for a decade.

He instigated and orchestrated the entire project at a hospital in Groningen, The Netherlands. He and three colleagues published a monograph in 1999, in the *British Medical Journal*. (Two co-authors, Ida Sabelis and Eduard Mooyaart, themselves engaged in intercourse in the MRI tube. Several other couples also contributed their all to the project.)

Called 'Magnetic Resonance Imaging of Male and Female Genitals During Coitus and Female Sexual Arousal', the study includes two copies of an MRI midsagittal image of 'the anatomy of sexual intercourse'. In the second copy, labels and hand-drawn outlines identify the bits that are of medical significance: ('P=penis, Ur=urethra, Pe=perineum, U=uterus, S=symphysis, B=bladder, I=intestine, L5=lumbar 5, Sc=scrotum').

Unknown to almost everyone, Dr van Andel asked the MRI technician to gather all the static images and assemble them together into a motion picture. The result: twenty-first century's greatest challenge to easy assumptions about porn.

For their illuminating report, van Andel, Willibrord Weij-mar Schultz, Eduard Mooyaart and Ida Sabelis were awarded the 2000 Ig Nobel Prize in medicine.

Schultz, Willibrord Weijmar, Pek van Andel, Ida Sabelis and Eduard Mooyaart (1999). 'Magnetic Resonance Imaging of Male and Female Genitals during Coitus and Female Sexual Arousal'. *British Medical Journal* 319 (18 December): 1596–1600.

DON'T LOOK

Dr Judith A. Reisman wants you to avoid looking at dirty pictures. Reisman wants you to look at her explanation of the horrible things dirty pictures can do to your brain, nervous system and civil rights. To make it easy for you to know what she is talking about, the good doctor has included some nice, dirty pictures in her report. To make it easy for you to read the report, she has put it on her website. You and your children can partake at www.drjudithreisman.org.

RESTRICTED TO ADULTS OVER AGE 18
SOME GRAPHIC IMAGES FROM MAINSTREAM PORNOGRAPHY

From 'The Psychopharmacology of Pictorial Pornography Restructuring Brain, Mind & Memory & Subverting Freedom of Speech'

The web site is rife with tributes to Reisman, beginning with one from Dr Laura Schlessinger, who says: 'Dr. Reisman has produced a scholarly and devastating study revealing the ugly and frighteningly dangerous pseudo-scientific assault on our children's innocence.' Schlessinger is herself renowned both for crusading against dirty pictures, and for the naked photographs of herself that are spread all over the Internet.

Reisman is president of the Institute for Media Education, in Granite Bay, California. She is sought worldwide, she says, 'to speak, lecture, testify, and counsel individuals, organiza-tions, professionals and governments regarding sex education and fraudulent sex scientists'. She has specifically dedicated herself to 'exposing Dr. Alfred C. Kinsey's fraudulent sex science research' in works such as *Kinsey, Sex and Fraud: The Indoctrination of a People*; *Kinsey: Crimes & Consequences* (now in its fourth edition); and, most recently, *Sexual Sabotage: How One Mad Scientist Unleashed a Plague of Corruption and Contagion on America.*

To do her work, Reisman has had to endure looking at a considerable amount of pornography. She describes some of it in her 1986 book *Images of Children, Crime and Violence in Playboy, Penthouse, and Hustler.* The book was subsidized with a grant from the US Department of Justice.

She later wrote a study called 'The Psychopharmacology of Pictorial Pornography Restructuring Brain, Mind & Memory & Subverting Freedom of Speech'. It carries a subtitle that is either warning or advertising, or perhaps both: 'Some Graphic Images from Mainstream Pornography'. The paper includes lots of pornography, alternating with dry technical drawings from neurobiology textbooks.

There is something for everyone, and lots of it. On page twenty-one, a detailed graphic depicting which brain regions have high concentrations of the neurotransmitter norepinephrine. A few pages later, a photo of a 'MEMORANDUM TO DR. JUDITH REISMAN FROM ANONYMOUS', for which this item is typical: 'Playboy, August 1975 – The bedroom, incestuous sadism exploited with the suggestion that we will be turned on by viewing a "hot" series of Jane (slightly exposed breasts and genitalia) in similar poses.'

A few of the porno pictures have white or black rectangles superimposed over the good bits. The neurobiology textbook diagrams are unretouched, as far as I could determine.

'The reason such a paper as this is necessary', Reisman writes, 'is due to the international inundation of sexual and sadosexual images and their direct, often fatal effect upon the conduct of millions of receivers of those images.'

Indeed. The paper is a morally bracing poke in the eye – just what the doctor (Dr Reisman) ordered. The lusty enjoyment with which it was written will not be lost on the reader.

Reisman, Judith A. (1990). *Kinsey, Sex and Fraud: The Indoctrination of a People*. Lafayette, LA: Huntington House.

— (1998). *Kinsey: Crimes & Consequences: The Red Queen & the Grand Scheme*. Arlington, VA: Institute for Media Education.

— (2010). *Sexual Sabotage: How One Mad Scientist Unleashed a Plague of Corruption and Contagion on America*. Washington, DC: WND Books.

— (1986). *Images of Children, Crime and Violence in Playboy, Penthouse, and Hustler*. Lafayette, LA: Huntington House.

— (2003). 'The Psychopharmacology of Pictorial Pornography Restructuring Brain, Mind & Memory & Subverting Freedom of Speech'. Working paper, Institute for Media Education.

WHEN SEX CAN BE AN EYE-OPENER

'Can Chlamydial Conjunctivitis Result from Direct Ejaculation into the Eye?' ask Doctors Simon Rackstraw, N.D. Viswalingam and Beng T. Goh of the Moorfields Eye Hospital in London. That question forms the title of a study they published in 2007 in the *International Journal of STD and AIDS*. In it, Rackstraw, Viswalingam and Goh describe the plights of four patients, and disclose the detective work involved in diagnosing and treating these unfortunate sufferers.

Conjunctivitis, whether of the variety caused by the sexually transmitted disease chlamydia or by some other, probably less colourful irritant, also goes by the name 'pink eye'. It's a catchall description for inflammation of the conjunctiva, the membranes that line both the eyelid and the surface of the eye itself.

The doctors finger the bacterium *Chlamydia trachomatis* as the likely culprit in as many as 9 percent of the severe conjunctivitis cases seen by casualty departments. They imply that most doctors would blindly – and perhaps wrongly – assume that manual transmission was involved: that patients had rubbed their own eyes with their own hands after those hands had come in contact with infected genital fluids, either their own or those of a spouse or a paramour.

One patient complained of 'a sore, red, sticky right eye; and a four-day history of stickiness in the left eye'. The others gave variations on the same theme.

The mechanical aspect of their tales, too, had a dull sameness.

One woman said that 'symptoms in her right eye started a week following sexual intercourse with a known male partner who had ejaculated into the eye'. Another explained that her 'partner had ejaculated into her right eye, following which she developed eye symptoms two weeks later'. A third 'recalled the eye symptoms starting two days after [her partner] ejaculated directly into her right eye'. The fourth patient, a man, 'had a casual encounter four months previously and recalled this partner ejaculating into his eye during oral sex'. The Sherlock Holmesian 'big clue', the study relates, was that 'none of the patients were found to have chlamydia detected within the genital tract on testing, but all gave a history of a recent sexual partner having ejaculated into the affected eye'. But the story remains incomplete, because 'Unfortunately, in these cases we were unable to test the sexual partners to see if they had chlamydia.'

Nonetheless, the four cases were eye-openers. Each patient, when questioned, coughed up that unexpected nugget of information. Each maintained, one way or another, that 'chlamydial conjunctivitis occurred following ejaculation of semen directly into the affected patients [sic] eye'.

The doctors reached a sort of conclusion, undoubtedly meant to guide the thoughts of any medical personnel who might have limited understanding as to the ways one can contract conjunctivitis. 'It is likely', they write, 'that this mode of transmission is underestimated as a history of ejaculation into the conjunctiva is not normally asked for'.

Rackstraw, Simon, N.D. Viswalingam and Beng T. Goh (2006). 'Can Chlamydial Conjunctivitis Result from Direct Ejaculation into the Eye?'. *International Journal of STD and AIDS* 17 (9): 639–41.

IN BRIEF

'FRACTURE PENIS: A CASE MORE HEARD ABOUT THAN SEEN IN GENERAL SURGICAL PRACTICE'

by Manash Ranjan Sahoo, Anil Kumar Nayak, Tapan Kumar Nayak and S. Anand (published in *BMJ Case Reports*, 2013)

The authors, at SCB Medical College, Cuttack, Odisha, India, begin their report in compelling fashion: 'A 36-year-old man presented to the emergency department with a history of trauma to genitalia during intercourse. The patient reported the forceful collision between his penis and the bed and audible clicking sound with swollen penis thereafter.'

MAY WE RECOMMEND

'A PENIS-SHORTENING DEVICE DESCRIBED BY THE 13TH CENTURY POET RUMI'

by C.W. Moeliker (published in the *Archives of Sexual Behavior*, 2007)

PLAYBOY PEER REVIEW

Personal peer review – peering at data with one's own eyes – is deemed crucial to some kinds of investigation. This eye-balled scrutiny figures in the war against indecent images of children. A study called 'Tanner Stage 4 Breast Development in Adults: Forensic Implications' looks at the very different things different experts saw when they all peered at the same female nipples.

Many websites with pictures of unclad persons feature a statement specifying that all those photographed are over eighteen. In some court cases, expert physicians testify as to the possible age of the people in some of those photos. The new study and other research demonstrates that such experts can be – and often are – wrong.

Many of those experts boast that they use a particular standard to judge the sexual maturity of a female body. That standard, called the Tanner Stages of Development or the Tanner Scale, was devised in the late 1960s by James M. Tanner and W.A. Marshall at the University of London's Institute of Child Health. It describes, in detail, how 192 white British girls' breasts and pubic hair changed in appearance as they became women.

Dr Arlan L. Rosenbloom, together with three colleagues at the University of Florida College of Medicine, undertook an experiment of sorts, relying solely on the Tanner Scale to suss out the ages of some women in photographs. They write: 'Inspired by the report of Italian and German investigators who used images from legitimate pornographic websites (to be sure that the subjects were women over 18 years of age), we examined 547 images with breast exposure from an anthology of the monthly centrefold illustrations in *Playboy* magazine from December 1953 to December 2007 that did not include more than one picture of any single model.' The team found that any doctor who relies on the Tanner Scale to judge such photos could mistakenly decide that about a quarter of them are too young to appear in such photographs.

In another study, in the *International Journal of Legal Medicine*, Rosenbloom documents in detail the 'high degree of inaccuracy' of medical expert testimony on this subject.

So, there is the problem that different doctors interpret (and misinterpret) the Tanner breast-development scale very (sometimes very, very) differently.

But there is a bigger problem: the scale does not measure age. It measures progression through 'apparent' signs of sexual development. Those appearances, and the ages at which they appear, can vary widely between individuals – which was part of the reason Tanner made the scale. Tanner himself wrote, in 1997, 'the Tanner Scales were not designed to be used for estimating chronological age, forensically or otherwise'.

The following year, Tanner and Rosenbloom together sent a letter about this to the journal *Pediatrics*, with the headline 'Misuse of Tanner Puberty Stages to Estimate Chronologic Age'. In response, other doctors wrote angry letters, each praising his own use of the Tanner Scale and denouncing Tanner.

Rosenbloom, Arlan L., Henry J. Rohrs, Michael J. Haller and Toree H. Malasanos (2012). 'Tanner Stage 4 Breast Development in Adults: Forensic Implications'. *Pediatrics* 130 (4): e978–81.

Marshall, W.A., and James M. Tanner (1969). 'Variations in Pattern of Pubertal Changes in Girls'. *Archives of Disease of Childhood* 44: 291–303.

Cattaneo, Cristina, Stefanie Ritz-Timme, Peter Gabriel, Daniele Gibelli, Elena Giudici, Pasquale Poppa, Doerte Nohrden, Sabine Assmann, Roland Schmitt and Marco Grandi (2009). 'The Difficult Issue of Age Assessment on Pedo-pornographic Material'. *Forensic Science International* 183 (1): e21–4.

Hefner, Hugh M., David Hickey, et al. (2008). *Playboy: The Complete Centerfolds.* San Francisco: Chronicle Books.

Rosenbloom, Arlan L. (2013). 'Inaccuracy of Age Assessment from Images of Postpubescent Subjects in Cases of Alleged Child Pornography'. *International Journal of Legal Medicine* 127 (2): 467–71.

Tanner, James M., and Arlan L. Rosenbloom (1998). 'Letter to the Editor: Misuse of Tanner Puberty Stages to Estimate Chronologic Age'. *Pediatrics* 102 (6): 1494.

RESEARCH SPOTLIGHT

'SEX DIFFERENCES IN RESPONSE TO RELATIONSHIP THREATS IN ENGLAND AND ROMANIA'
by Gary L. Brase, Dan V. Caprar, and Martin Voracek (published in the *Journal of Social and Personal Relationships*, 2004)

ISLANDS OF INTEREST

Sex clearly drives Britain and Argentina as they vie to dominate islands of interest. The two great nations are rivals in producing academic studies of whether and how people stare at women's breasts or buttocks.

Britain fired the first shot in this war. In 2007, Adrian Furnham and Viren Swami of University College London published a report called 'Perception of Female Buttocks and Breast Size in Profile', in the journal *Social Behaviour and Personality*.

Professor Furnham is, by his own reckoning, one of the most productive academics alive, publishing many hundreds of papers in dozens of far-flung fields. Professor Swami, at the University of Westminster since 2007 and an adjunct reader in psychology at HELP University College in Kuala Lumpur, also beavers relentlessly, both in Britain and abroad. He has launched treatises in topics from the physical attractiveness of women and men in London's boroughs (see page 297) to the opinions of Austrian students as to the personalities of butchers and hunters.

The Furnham/Swami breast/buttocks paper gathers together many facts known only to a small number of specialists. Here's one. Furnham and Swami write: 'It is widely recognised, for example, that the African Hottentot (Caboid) tribe and certain tribes in the Andaman Islands show a preference for large fat deposits on the buttocks, a condition known as steatopygia. It has been suggested that such fat deposits on the buttocks and thighs may signal resource accrual...'

Furnham and Swami studied the reactions of 114 British undergraduates – men and women. They write: 'The stimuli consisted of nine nude female silhouettes, prepared ... in such a manner that the size of breasts and buttocks could be varied systematically.' Their conclusion: 'The participants in this study showed a preference for small breast size, although buttocks size did not appear to alter ratings of attractiveness. It would be useful for future research to include a larger range of breast sizes and shapes ... to investigate the possibility that optimal breast size ... varies across individuals'.

Argentina took five years to respond to this provocation.

Mariano Sigman, director of the Laboratory for Integrative Neuroscience at the University of Buenos Aires, and two collaborators concentrated their analytical firepower on male undergraduates. Their study called 'Eye Fixations Indicate Men's Preference for Female Breasts or Buttocks' was published in the *Archives of Sexual Behaviour* in 2012.

The Sigman team reports that:

1) 'Argentinian males tended to define themselves as favouring breasts or buttocks';
2) 'the distribution was biased towards buttocks'; and
3) individuals who say they prefer to gaze at one of those body parts behave as if they do.

The British and Argentinian projects both exhibit indomitable focus. The authors and test subjects apparently refuse to let anything seriously distract them from what they want to study.

Furnham, Adrian, and Viren Swami (2007). 'Perception of Female Buttocks and Breast Size in Profile'. *Social Behavior and Personality* 35 (1): 1–8.
Dagnino, Bruno, Joaquín Navajas and Mariano Sigman (2012). 'Eye Fixations Indicate Men's Preference for Female Breasts or Buttocks'. *Archives of Sexual Behavior* 41 (4): 929–37.

IN BRIEF

'EFFECT OF DIFFERENT TYPES OF TEXTILES ON MALE SEXUAL ACTIVITY'
by A. Shafik (published in *Archives of Andrology*, 1996)

The author explains: 'Each of the 4 test groups were dressed in one type of textile underpants made of either 100% polyester, 50/50% polyester/cotton mix, 100% cotton, or 100% wool. Sexual behavior was assessed before and after 6 and 12 months of wearing the pants, and 6 months after their removal ... [and] ... polyester underpants could have an injurious effect on human sexual activity.'

MAY WE RECOMMEND

'Real men wear kilts'. The anecdotal
evidence that wearing a Scottish kilt has
influence on reproductive potential: how
much is true?

EJO Kompanje

Abstract
Background and aims: There are anecdotal reports that men who wear (Scottish) kilts have better sperm quality and
better fertility. But how much is true? Total sperm count and sperm concentration reflect semen quality and male
reproductive potential. It has been proven that changes in the scrotal temperature affect spermatogenesis. We can ar

'REAL MEN WEAR KILTS: THE ANECDOTAL EVIDENCE THAT WEARING A SCOTTISH KILT HAS INFLUENCE ON REPRODUCTIVE POTENTIAL — HOW MUCH IS TRUE?'

by Erwin J.O. Kompanje (published in the *Scottish Medical Journal*, 2013)

A HISTORY OF WORDPLAY

Words, words, words are the bread, butter, salt, pepper, meat and potatoes of a small, US-based magazine called *Word Ways* that has been coming out four times a year since 1968. Dmitri Borgmann, the founding editor, described it as 'the journal of recreational linguistics'. Its essence, in a word: wordplay.

Borgmann's obituary, in a 1985 issue of *Word Ways*, says his greatest achievement was to 'demonstrate that wordplay is an intellectual discipline in its own right'. Borgmann's reputation was already such, says the obituary, that Standard Oil of New Jersey had hired him to devise a replacement for its antiquated brand name. 'Twas Borgmann, they say, who spiffed and twisted old-fashioned 'Esso' into modern 'Exxon'. (Later issues of *Word Ways* say that the Esso-into-Exxon story may be rather more complicated.)

The first issue of *Word Ways* included Borgmann's 'The Longest Word', in which he traipses along the length of 'the

27-letter honorjficajhlitudinitatibus', 'the 28-letter antidisestab-lishmentarianism', 'the 45-letter pneumonoultramicroscopic-silicovolcanokoniosis', and on eventually to a chemical name that is 1,913 letters long. The first two hundred of those letters, by the way, are

> metrionylglutaminylarginyltyrosylglutamylseryl-leucylphenylalanylalanylglutaminylleucyllysylgluta-mylarginyllysylglutamylglycylalanylphenylalanylva-lylprolylphenylalanylvalylthreonylleucyl glycylaspartyl

A 1986 article called 'Dr. Awkward and Olson in Oslo' by Lawrence Levine begins: 'The long voyage between my first tentative effort at constructing a short palindrome of some 40 letters, and the eventual completion of a palindromic novel numbering 31,594 words (or approximately 104,000 letters) some 20 years later, was an unrelenting lesson in many dis-ciplines. There were lessons in trial and error ...' Levine care-fully states his creed as a creator of these bidirectional hunks of text: 'One must not cheat by inventing words or coining new spellings.'

In 1995, a pseudonymous author wrote a short survey of 'alphabet poems' – poems in which each line is keyed to a letter of the alphabet. Many were intended, directly or indirectly (being read aloud) for young children. The article presents one, with authorship attributed to Rod Campbell in 1988, which it calls 'uninspired', and which begins:

> *a is for apple ready to eat*
> *b is for boots to put on your feet ...*

At the other reach of complexity comes a poem written in the late 1800s, authorship uncertain, about the Crimean war. It begins:

> *An Austrian army, awfully arrayed*
> *Boldly by battery besieged Belgrade ...*

In 1991, Darryl Francis of Mitcham, Surrey, took an earthy look at his native land, in an essay called 'Naughty Words in British Placenames'. Francis claimed to be restraining himself by giving only sixty-nine examples, though he was rather unrestrained in his view of the boundaries of Britain. Here are six of Francis's finds, with his own distinctive capitalization and description:

> **ARSEnal**: an area of the London borough of Islington, and also the name of a railway station in the area
> **BREAST**: an island in County Wexford, Ireland
> **BUMble hole**: a locality in Worscestershire
> **prettyBUSH**: a locality in County Wicklow, Ireland
> **CRAPstone**: a hamlet in Devon
> **PEninNIS**: a pile of rocks in the Scilly Isles
> **PRATt's Bottom**: a hamlet in Kent
> **SHITlington**: a parish in West Yorkshire
> **dURINEmast**: a loch in Argyllshire
> **WILLY**: a parish in Northamptonshire

n.a. (1985). 'Dmitri Borgmann, Father of Logology'. *Word Ways: The Journal of Recreational Linguistics* 18 (1): 3–5.

n.a. (2012). 'Colloquy'. *Word Ways: The Journal of Recreational Linguistics* 24 (3): 151–3.

Borgmann, Dmitri (1968). 'The Longest Word'. *Word Ways: The Journal of Recreational Linguistics* 1 (1): 33–5.

Levine, Lawrence (2011). 'Dr. Awkward and Olson in Oslo'. *Word Ways: The Journal of Recreational Linguistics* 19 (3): 140–5.

Indicator, Nyr (2012). 'Alphabet Poems: A Brief History'. *Word Ways: The Journal of Recreational Linguistics* 28 (3): 131–5.

Francis, Darryl (1991). 'Naughty Words in British Placenames'. *Word Ways: The Journal of Recreational Linguistics* 24 (2): 84–6.

MAY WE RECOMMEND

'POSITION PAPER: MANAGEMENT OF MEN COMPLAINING OF A SMALL PENIS DESPITE AN ACTUALLY NORMAL SIZE'
by Hussein Ghanem, Sidney Glina, Pierre Assalian and Jacques Buvat (published in *Journal of Sexual Medicine*, 2013)

MONKEYING AROUND

The authors of a study called 'High Frequency of Postcoital Penis Cleaning in Budongo Chimpanzees' do not beat about the bush. 'We report on postcoital penis cleaning in chimpanzees', they write. 'In penis cleaning, leaves are employed as "napkins" to wipe clean the penis after sex. Alternatively, the same cleaning motion can be done without leaves, simply using the fingers. Not all chimpanzee communities studied across Africa clean their penes and, where documented, the behaviour is rare. By contrast, we identify postcoital penis cleaning in Budongo Forest, Uganda, as customary.'

Sean O'Hara, a Durham University anthropologist (who has since moved to the University of Salford), and Phyllis Lee, a psychology professor at the University of Stirling, published their monograph in the journal *Folia Primatologica* in 2006. The British team lists the few instances in which humans had documented the practice. Jane Goodall 'mentions it in the Gombe chimpanzees, Tanzania, and leaf napkin use in Kibale forest, Uganda, is known ... and in 25 years of observation at Taï Forest, Côte d'Ivoire, 'leaf-wipe' has been recorded just once'.

O'Hara and a field assistant named Monday Gideon did the Budongo detecting 'between January and September 2003 and were able to verify "cleaning" or "not cleaning" for 116 copulations. Penis cleaning occurred in 34.5% of copulations (9.5% with leaf napkins and 25% without use of a tool)'. The team expresses wonder that this particular form of tool use varies so starkly in popularity. 'For penis wiping to be common in some locations while rare or absent elsewhere presents a puzzle', they say.

They point out that many kinds of animals use one or another type of tool. They cite reports about New Caledonian crows, bottlenose dolphins, parasitoid wasps, capuchin monkeys, and other species. O'Hara and Lee explain that most of these tool-using practices are 'cultural behaviors' – that is, learned from fellow dolphins, wasps, monkeys, or whatever. What's especially notable here, they say, is that 'few material cultural behaviours are conducted in asocial contexts … Postcoital penis cleaning is one such activity. Although the copulatory act is, by definition, a social event encompassing more than one individual, the penis wiping that follows is solitary and self-directed'.

They note the existence of hypotheses that the cleaning serves some important, particular function. The males do it to check for signs of sexually transmitted disease [STD], perhaps, or maybe to monitor some reproductive aspect of the females with whom they consort.

But O'Hara and Lee keep a disciplined focus on the main question: culture. 'Whatever the motivation or function', they write, 'Budongo males appear more fastidious in penis hygiene than elsewhere. We found no proclivity for the use of specific leaf types; leaves appeared to be plucked non-systematically … While the functional or STD context remains unclear, we suggest that using leaf napkins is a cultural trait in chimpanzees.'

O'Hara, Sean J., and Phyllis C. Lee (2006). 'High Frequency of Postcoital Penis Cleaning in Budongo Chimpanzees'. *Folia Primatologica* 77 (5): 353–8.

HAVING THE TIME OF YOUR MONTH

The 'Menstrual Joy Questionnaire' was developed in 1987. It entered the world as part of a book called *The Curse: A Cultural History of Menstruation*, written by Janice Delaney, Mary Jane Lupton and Emily Toth. They were distressed at the existence and influence of the 'Menstrual Distress Questionnaire', a dour piece of work created nineteen years earlier by Rudolf H. Moos at Stanford University.

Moos was a psychiatrist. He delved, professionally, into many kinds of distress, among them: depression, problem drinking, work-induced stress and the social atmospheres of psychiatric wards. Though few held it against him, Moos had little first-person experience of menstrual emotions. His was a rigorous academic understanding.

The three menstrual joy scholars were a cheerier lot. They were literary folk. Delaney was director of a prestigious fiction-writing award given by the Folger Library in Washington DC. Lupton and Toth were English professors: Lupton at Morgan State University in Baltimore, Toth at Pennsylvania State University. Their menstrual savvy came from personal experience supplemented by a vast knowledge of literature.

The Menstrual Joy Questionnaire is short and simple, inquiring into ten joyful menstrual matters, specifically: (1) high spirits; (2) increased sexual desire; (3) vibrant activity; (4) revolutionary zeal; (5) intense concentration; (6) feelings of affection; (7) self-confidence; (8) feelings of euphoria; (9) creativity; and (10) feelings of power.

Seven years after Delaney, Lupton and Toth launched their admittedly whimsical questionnaire, a team of researchers tried to gauge its impact. Joan Chrisler, Ingrid Johnston, Nicole Champagne and Kathleen Preston of Connecticut College published a study, called 'Menstrual Joy: The Construct and Its Consequences', in the journal *Psychology of Women Quarterly*. Their purpose, they stated, was 'to examine participants' reactions to the concept of menstrual joy … We found it too difficult to resist the temptation to see what women would think of the construct.'

And so they gave the Menstrual Joy Questionnaire to forty women. Then they asked five questions:

a) What was your reaction to seeing a questionnaire entitled 'Menstrual Joy'?

b) Have you previously regarded menstruation as a positive event in your life? If yes, describe the menstrual cycle's positive aspects in your own words.

c) Did the Menstrual Joy Questionnaire encourage you to view menstruation in a different way? If yes, please explain.

d) Do you think you will be aware of or anticipate some of these positive aspects during your next menstrual cycle?

e) Do you discuss menstruation openly? If so, with whom?

Here, in the researchers' own words, is what they learned: 'The most common reactions to the questionnaire were incredulity or disbelief (27.5%), shock or surprise (22.5%) or the belief that the title was sarcastic or ironic (25%). Other participants expressed initial interest (12.5%), amusement (12.5%), confusion (12.5%), irritation or annoyance (5%), appreciation (2.5%), or sadness (2.5%). Some participants expressed more than one reaction.'

'The results of this study', they concluded, 'are interesting for several reasons.'

Several years later, two British psychologists, Aimee Aubeeluck at the University of Derby and Moira Maguire at the University of Luton, decided to replicate Chrisler et al's experiment, but chose to remove the title of the questionnaire altogether, so as not to influence a woman's reaction by hinting at either 'joy' or 'distress'. They say that the wording of 'Joy' questions alone was enough to make women think more favourably about menstruation 'as a natural event'.

Delaney, Janice, Mary Jane Lupton and Emily Toth (1987). *The Curse: A Cultural History of Menstruation.* Chicago: University of Illinois Press.

Moos, Rudolf H. (1968). 'The Development of a Menstrual Distress Questionnaire'. *Psychosomatic Medicine* 30: 853–67.

Chrisler, Joan C., Ingrid K. Johnston, Nicole M. Champagne and Kathleen E. Preston (1994). 'Menstrual Joy: The Construct and Its Consequences'. *Psychology of Women Quarterly* 18 (3): 375–87.

Aubeeluck, Aimee, and Moira Maguire (2002). 'The Menstrual Joy Questionnaire Items Alone Can Positively Prime Reporting of Menstrual Attitudes and Symptoms'. *Psychology of Women Quarterly* 26 (2): 160–2.

RESEARCH SPOTLIGHT

'MOOD AND THE MENSTRUAL CYCLE: A REVIEW OF PRO-SPECTIVE DATA STUDIES'

by Sarah Romans, Rose Clarkson, Gillian Einstein, Michele Petrovic and Donna Stewart (published in *Gender Medicine*, 2012)

'THE GENDERED OVARY: WHOLE BODY EFFECTS OF OO-PHORECTOMY'

by Gillian Einstein, April S. Au, Jason Klemensberg, Elizabeth M. Shin and Nicole Pun (published in the *Canadian Journal of Nursing Research*, 2012)

'THE HERMUNCULUS: WHAT IS KNOWN ABOUT THE REPRESENTATION OF THE FEMALE BODY IN THE BRAIN?'

by Paula M. Di Noto, Leorra Newman, Shelley Wall and Gillian Einstein (published in the *Cerebral Cortex*, 2012)

> **The *Hermunculus*: What Is Known about the Representation of the Female Body in the Brain?**
>
> Paula M. Di Noto[1], Leorra Newman[1], Shelley Wall[2] and Gillian Einstein[1,3]
>
> [1]Department of Psychology and, [2]Biomedical Communications, Institute of Medical Science and [3]Dalla Lana School of Public Health, University of Toronto, Toronto, Ontario, Canada

EXPONENTS OF ARTISTS' SEXUAL ACTIVITIES

To deal with their realization that some artists get a lot of sex while others get little or none, Helen Clegg, Daniel Nettle and Dorothy Miell made use of an ancient tool – a tool that mathematicians count among the sexiest of mankind's inventions. The logarithm.

The trio had joined forces, as they later described it, to 'investigate the relationship between mating success and artistic success in a sample of 236 visual artists'.

Clegg is a University of Northampton senior lecturer in psychology, Nettle a professor of behavioural science at Newcastle University and Miell the head of the College of Humanities and Social Science at the University of Edinburgh. Their report, called

'Status and Mating Success Amongst Visual Artists', appears in the journal *Frontiers in Psychology*.

The study gives us barely any numerical detail. It says only this: 'The distribution of number of sexual partners for these participants was highly skewed with a minimum of 0 and a maximum of 250 (M=10.67). Therefore, the data were converted to a log scale and [we performed our analysis] using this scale.'

That 'M=10.67' is the median. Half of the 236 artists had had, each of them, fewer than 10.67 lovers. The other artists each had had in excess of 10.67 bed-mates. Or so they told the researchers.

Two lovers. Twenty lovers. Two hundred lovers. They seem almost to be from different universes, the collections of five or six lovers, versus the serial harems of one hundred or two hundred. How to talk coherently about a hodgepodge of small and big numbers? You do it with logarithms.

Roughly speaking (I don't have room here to go into much detail), the logarithm of a particular number tells – measures, really – how many extra digits that number has. The number 1 has no extra digits. Its logarithm is zero. The number 10 has one extra digit. Its logarithm is 1. The number 100 has two extra digits; its logarithm is 2. The logarithm of 101 is ever-so-slightly bigger than 2 (it's about 2.0043). The logarithm of 250 is bigger still (about 2.3979).

The logarithm is a concise, rough way to compare things across vast scales of bigness and smallness. That painter who's got a new girlfriend every few months? About log 2. That lonely graffiti gal whom everyone shuns? Log 0, it seems.

The researchers used logarithms also when they tried to understand a related set of numbers. They had computed what they call the 'mating strategy index' of the various artists. 'Each one-night stand gained one point, each relationship up to a month two points, and soon up to each relationship 10 years or over, which gained eight points. The total number of points for each person was added up and divided by their total number of relationships.'

After tiptoeing through all their data and computations, the artists-and-sex researchers decided that 'more successful male artists had more sexual partners than less successful artists, but this did not hold for female artists'.

Clegg, Helen, Daniel Nettle and Dorothy Miell (2011). 'Status and Mating Success amongst Visual Artists'. *Frontiers in Psychology* 2 (October): 310.

AN IMPROBABLE INNOVATION

'PRODUCING REPLICAS OF BODY PARTS'
a/k/a a moulding and cloning machine, by George Castanis and Thaddeus Castanis (US patent no. 4,335,067, granted 1982)

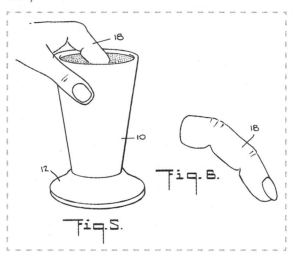

Form and cast finger, suitable for 'simple cloning operations, particularly in the hands of children'

NAUGHTY THOUGHTS, HEMISPHERICALLY

When a person thinks about naughty things, does one side of the brain get more exercised than the other? Eight scientists studied that question. Their report, 'Hemispheric Asymmetries during Processing of Immoral Stimuli', appears in the journal *Frontiers in Evolutionary Neuroscience*. The stated goal is to describe 'the neural organization of moral processing'.

Debra Lieberman, a professor of evolutionary psychology at the University of Miami, Florida, acts as spokesperson for the team. Other members are based at Miami, at the University of New Mexico and at Stanford University in California. Another, Walter Sinnott-Armstrong, is at Duke University in North Carolina.

They had to work with a few limitations – the same limitations that apply to anyone who tries to describe what's going on in the brain.

With the exception of a few crackpots or geniuses, scientists don't claim to understand how the 100,000,000,000 or so parts of the human brain manage to think thoughts. Many of those multitudinous parts are connected to one another in complex ways that are quirkily different in every person. Some of the connections change over the course of a life, or a day, or even a few minutes. Many tiny brain parts are clumped into big conglomerations, some quite distinct (hello, cerebellum!), but others have fuzzy locations and borders.

The study does not risk getting bogged down in those larger, complicated conundrums. It restricts itself to the simple question: how does immorality play out in the brain?

The scientists sought their answer by recruiting some test subjects. They confronted each volunteer with several levels of immorality, in the form of words and images.

The team used MRI machines to monitor indirectly (via electromagnetic emissions) where largish amounts of blood flowed in the brain as each volunteer confronted each example of immorality. In theory, anyway, blood flows most freely near whichever brain parts are actively thinking, or have just thought, or are just about to think, or are busily doing something else.

In one test, volunteers saw different kinds of printed statements. Some were about pathogens ('You eating your sister's spoiled hamburger, You sipping your sister's urine, You eating your sister's scab'); some about incest ('You giving your sister an orgasm, You watching your sister masturbate, You fondling your sister's nipples'); some about 'nonsexual immoral acts' ('You burgling your sister's home, You killing your sister's child'); and others about 'neutral acts' ('You reading to your sister, You holding your sister's groceries').

> ***Stimuli and task***
> During scanning, participants were given a simple memory-recognition task that has been shown previously to tap affective processing (Kiehl et al., 2005). Participants saw a total of four types of statements: statements involving pathogen-related acts (Pathogen; e.g., You eating your sister's spoiled hamburger, You sipping your sister's urine, You eating your sister's scab), incestuous acts (Incest; e.g., You giving your sister an orgasm, You watching your sister masturbate, You fondling your sister's nipples), non-sexual immoral acts (Non-Sexual Immoral; e.g., You burglarizing your sister's home, You killing your sister's child, You knocking your sister down the stairs), and neutral acts (Neutral; e.g., You reading to your sister, You holding your sister's groceries, You closing the door to your sister's closet). Statements were divided into

'Stimuli and task' of 'Hemispheric Asymmetries during Processing of Immoral Stimuli'

In other tests, volunteers saw other kinds of statements or pictures, each chosen for its evident moral content.

After all the immorality was seen, and the measurements made, the researchers calculated that the left side of the brain had been more involved than the right side. Thus, concludes the study: 'There is a left-hemisphere bias for the processing of immoral stimuli across multiple domains.'

Cope, Lora M., Jana Schaich Borg, Carla L. Harenski, Walter Sinnott-Armstrong, Debra Lieberman, Prashanth K. Nyalakanti, Vince D. Calhoun and Kent A. Kiehl (2010). 'Hemispheric Asymmetries during Processing of Immoral Stimuli'. *Frontiers in Evolutionary Neuroscience* 2 (110): 1–14.

Lieberman, D., and D. Symons (1998). 'Sibling Incest Avoidance: From Westermarck to Wolf'. *Quarterly Review of Biology* 73 (4): 463–6.

Schaich Borg, J., D. Lieberman and K.A. Kiehl (2008). 'Infection, Incest, and Iniquity: Investigating the Neural Correlates of Disgust and Morality'. *Journal of Cognitive Neuroscience* 20: 1529–46.

Lieberman, D., D.M.T. Fessler and A. Smith (2011). 'The Relationship between Familial Resemblance and Sexual Attraction: An Update on Westermarck, Freud, and the Incest Taboo'. *Personality and Social Psychology Bulletin* 37: 1229–32.

DeBruine, L.M., B.C. Jones, J.M. Tybur, D. Lieberman and V. Griskevicius (2010). 'Women's Preferences for Masculinity in Male Faces Are Predicted by Pathogen Disgust, But Not Moral or Sexual Disgust'. *Evolution and Human Behavior* 31: 69–74.

MAY WE RECOMMEND

'THE "SMELLSCAPE" OF MOTHER'S BREAST: EFFECTS OF ODOR MASKING AND SELECTIVE UNMASKING ON NEONATAL AROUSAL, ORAL AND VISUAL RESPONSE'

by Robert Soussignan, Paul Sagot and Benoist Schaal
(published in *Developmental Psychobiology*, 2007)

KERBSTONE COPS

Policewomen who work undercover as prostitutes have certain needs. Scholars have not addressed those needs until recently. Or at least, no scholar has done so in a way that would be accepted for formal publication in a research journal in the field of information science.

Lynda M. Baker is an associate professor in the library and information science programme at Wayne State University in Detroit, Michigan. Her study, called 'The Information Needs of Female Police Officers Involved in Undercover Prostitution Work', was published in the journal *Information Research*.

Baker interviewed seven vice officers, and then observed two of them in action. She learned that 'officers need a variety of information'. One decoy officer refused to co-operate in the study; Baker does not inform us as to why.

The profession, like all professions, has its own peculiar information requisites. Baker writes that: 'To be a credible decoy, officers need to know and become comfortable with the language of the street. Some officers said they learned it on the street by listening to real prostitutes. Others consulted fellow officers for clarification of unknown terms. One officer mentioned receiving a booklet from the police department in Las Vegas with updates on changes in terms. The crucial point is that, to make their case, the officers need to understand completely what a john is requesting. If they are conversing with a john who is using strange terms, Officers C and G stated that they feign ignorance by stating: "I'm from [another city] and we don't use that in [name of the city]. What're you talking about?" '

A decoy exchanges information with four groups of people: johns, prostitutes, vice squad team members and members of the community. Each calls for a different type of what Professor Baker calls 'information behavior'. This is further specialized for each region of the city.

The officers need to 'know what to wear and how to act. If they are in a drug-infested, low-paying area, the decoys will wear older clothes and shoes, may blacken a tooth, and apply makeup haphazardly … To make their case, the decoys must seek information from the johns, that is, a request for a sex act in exchange for something of value.'

Baker points out that 'Information is both given to and sought from prostitutes who are working in the area of the decoy operation. Because the prostitutes can be territorial, telling them, "This is my corner", often works for the decoys.'

Information does flow in two directions, and it is valuable. Accounting for that can be tricky, points out Baker: 'One business owner, who did not know that a decoy operation was underway in front of his store, called the police on a decoy. On the street, therefore, the decoy walks a tight line when it comes to informing business owners about her work.'

In a separate report Baker assessed the attitudes and experiences of seven police decoys – presumably the same seven. She reports that 'Most of the officers described their work as interesting'.

Baker, Lynda M. (2004). 'The Information Needs of Female Police Officers Involved in Undercover Prostitution Work'. *Information Research* 10 (1): n.p., http://informationr.net/ir/10-1/paper209.html.
— (2007). 'Undercover as Sex Workers: The Attitudes and Experiences of Female Vice Officers'. *Women & Criminal Justice* 16 (4): 25–41.

IN BRIEF

'THE STRESSFUL KISS: A BIOPSYCHOSOCIAL EVALUATION OF THE ORIGINS, EVOLUTION, AND SOCIETAL SIGNIFICANCE OF VAMPIRISM'

by Donald R. Morse (published in *Stress Medicine*, 1993)

The author, at the Temple University School of Dentistry, in Philadelphia, explains: 'On the positive side, vampirism can provide temporary escape from the stressors of the 1990s; on the negative side is the sinister nature of engaging in ritualistic, cultic vampirism.'

EIGHT

WHAT COMES AFTER THE COLON?

IN BRIEF

'PRESSURES PRODUCED WHEN PENGUINS POOH: CALCULA-TIONS ON AVIAN DEFAECATION'

by Victor Benno Meyer-Rochow and Jozsef Gal (published in *Polar Biology*, 2003, and honoured with the 2005 Ig Nobel Prize in fluid dynamics)

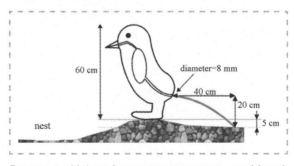

Penguano, with 'rectal pressure necessary to expel faecal material over a distance of 40 cm'

Some of what's in this chapter: Some – but not all – call it waste • Cucumbers sans burp • Personal ammunition, within and then without tanks • Bean outcome surprise • Gods' pollution • Addicted to cow dung fumes • What relative delight comes from your child • Child-parent feedback • How disgusting, please? • Eminence in rat-catching • Dog-savoured dreck • Cat-box smell-test chamber • Unbuild a tower • Finger-skills for nurses

TALKING RUBBISH

When economists talk economics, some of them talk rubbish. Few mean it as plainly, as directly, as Alexi Savov. Savov wrote a study called 'Asset Pricing with Garbage', which filled twenty-four pages of the *Journal of Finance* early in 2011.

To Savov, garbage is valuable not only for its own worth, but because, in a mathematical sense, it represents many of the things that people and corporations treasure most. Maybe, just maybe, he implies, the rises and falls in garbage production reliably and fairly accurately measure what a society is worth. Savov, an assistant professor of finance at New York University's Leonard N. Stern School of Business, did the garbage research a few years earlier when he was still at the University of Chicago.

Economists struggle, always, to get a better mental grasp of the messy confusion known as 'the economy'. Some economists are consumed with the economic concept called 'consumption'. They want to know how much stuff – solids, liquids, gases, energy, services, whatever – get consumed during different years.

But these economists disagree violently about which stuff to measure. Savov's garbage work takes its place in the long line of studies wrestling with the worth-versus-worthlessness of measuring all sorts of durable goods (cars, kettles), private goods (chocolate bars, gift copies of *Fifty Shades of Grey*), public goods (roads, statues of Margaret Thatcher), luxury goods (yachts, diamond bling), energy, services and whatnot.

Savov says he analysed '47 years of annual data from the US Environmental Protection Agency (EPA) … I use municipal solid waste (MSW), or simply garbage, as a new measure of consumption. Virtually all forms of consumption produce waste, and they do so at the time of consumption. Rates of garbage generation should be informative about rates of consumption.'

Savovian garbage includes detritus from both homes and businesses. 'Everyday items such as product packaging, grass clippings, furniture, clothes, bottles, food scraps, newspapers, appliances and batteries.' It excludes materials, typified by construction waste and municipal wastewater treatment sludge, that are sent directly to landfills.

Savov checked his methods by applying them also to ten years of garbage data from nineteen European countries including the UK. He found the Euro-garbage econometric performance to be 'consistent with the US results'.

His paper points out many subtleties in the relationship of garbage to things that his profession has traditionally tracked and esteemed – luxury goods, stocks and bonds – as indicators of the worth of our wealth. Garbage, he concludes, gives a solider, less often illusory, picture of the economy.

The final sentence of Savov's study adds meaning to the old saying 'garbage in, garbage out'. Savov writes: 'The relative success of garbage as an alternative measure of consumption raises the possibility that the failure of the standard consumption-based model is due to a failure to measure consumption properly.'

Savov, Alexi (2011). 'Asset Pricing with Garbage'. *Journal of Finance* 66 (1): 177–201.

QUESTIONING BURPING

'What Are Burpless Cucumbers?' This apparently simple question is perplexing to many of the people who produce and sell the green, oblong objects. Todd C. Wehner, a professor of horticultural science at North Carolina State University, tried to clarify the matter by conducting an experiment. He fed burpless cucumbers to both burpable and burpless judges, then published his results in a study called 'What Are Burpless Cucumbers?'

The marketplace offers conflicting information about burpless cucumbers. Some seed purveyors claim the word 'burpless' is just a synonym for 'seedless'. Others, the Burpee seed company among them, blithely fudge on the matter. Burpee offers eight kinds of burpless cucumber, including one called Big Burpless Hybrid. Burpee says: 'These varieties have almost no seed cavities for easier digestion.'

> **Methods and materials**
>
> BURPINESS TESTING. Three cultivars were used: Marketmore 76 (normal-bitter), Marketmore 80 (bitterfree), and Tasty Bright (burpless). Seeds were obtained from Dr. Henry Munger, Cornell University, Ithaca, NY ('Marketmore 76' and 'Marketmore 80'), or Sakata Seed America ('Tasty Bright'). Fruit from seedlings were rated for bitterness as described in the next section.
>
> Burpiness was evaluated using field-grown cucumbers. The experiment was a split plot with two replications and two harvests. Whole plots were two seasons (spring and summer within 1 year), subplots were the three cultivars (Marketmore 76, Marketmore

Methods and materials for 'Burpiness Testing'

But some are confident about the vegetable's burplessness.

The Thompson & Morgan seed catalogue, for example, offers (at the time of writing) the Burpless Tasty Green F1 variety at a price of £2.49 for a packet of ten seeds, with this assurance:

Crops in: 58 Days

Description: Yes, it's true. No indigestion problems
with this cucumber. Flavour is superb, crisp and
delicious – anyone can eat it and they are very
easy to grow.

Given this well-seeded disagreement and confusion, the community of cuke farmers, gardeners and consumers suffered. Todd C. Wehner saw the confusion, and acted.

The main objective of Wehner's research, he writes in the rather awkwardly named journal *Hortechnology*, 'was to determine whether oriental trellis cucumbers cause less burping when eaten'.

Wehner specializes in the study of cucurbits – the plant family whose siblings include cucumber, watermelon, cantaloupe and certain gourds. He is an acknowledged expert on watermelon DNA, on sex expression in luffa gourds, and on other matters that, throughout most of history, were utter mysteries. He took a straightforward approach to this new question.

'It has been suggested by researchers that burpless cucumbers contain less of a burping compound … or are just the marketing term for oriental trellis cucumbers', he writes. 'The objective of this experiment is to determine whether oriental trellis cucumbers cause less burping when eaten.'

Wehner fed three kinds of cucumber to six judges. 'Judges were grouped (three each) into susceptible or resistant, based on their previous experience with cucumbers. Fruits were evaluated for burpiness using six judges eating a 100-millimeter length of fruit per day. Burpiness was measured on a 0 to 9 scale … Ratings of burpiness were made within an hour of eating.' The trials went on for three days. Upon sifting through his data, Wehner discovered that the judges who were susceptible to burping burped slightly less after eating the burpless cucumbers.

But this may not be the end of the story. Professor Wehner takes care to point out that 'Additional research is needed on cucumbers of all types to identify cultivars that are free of burping for susceptible judges, and to identify the compound responsible for the burping effect.'

Wehner, Todd. C. (2000). 'What Are Burpless Cucumbers?'. *Hortechnology* 10 (2): 317–20.

RESEARCH SPOTLIGHT

'QUEASINESS: AN INFORMAL LOOK'
by John W. Trinkaus (published in *Perceptual and Motor Skills*, 1990)

LAYING WASTE ONTO THE ENEMY

Aleksandr Georgievich Semenov patented an efficiently disgusting weapon system. Using his method, soldiers inside an armoured tank, under battle conditions, can dispose of their biological waste products in an unwasteful way: encasing those materials, together with explosives, in artillery shells that they then fire at the enemy.

Semenov, residing in St Petersburg, can and does brag of having Russian patent no. 2,399,858, granted in 2009, officially titled 'Method of Biowaste Removal from Isolated Dwelling Compartment of Military Facility and Device for Its Implementation'.

As patents go, it's of modest length: twelve pages, with only two technical drawings. The original document is mostly in Russian. The prolific inventor (he has about two hundred other patents), sent me a full translation into English.

Figure 2 seems, at a glance, unremarkable: a cutaway side view of an artillery shell in its cartridge, tipped with a screw-in 'nose cone'. There's a large charge to (in Semenov's description) 'burst' the shell, and a small charge to trigger that burst. The shell itself contains – indeed is mostly – a compartment for the payload.

Figure 1 is larger and more complex, showing the entire tank in cutaway side view. A single crewman perches inside. Beneath him, an empty shell collects the waste that emerges from his anus.

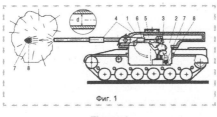

Figure 1

Here's how the patent describes the scene, with numbers referring to specific items in the drawing: 'A military man (3) puts the wastes (8) into the capacity (7) directly (fig.1) or in two steps. After it's complete or, if it is necessary, incomplete filling the capacity is tightly sealed by the cover.' The key action is in one sentence: 'The gun charged with special projectile is targeted on a safety zone or on any enemy target which is worth for catching it.'

As the projectile leaves the tank, it removes what would eventually have become a source of stinking misery for the poor soldiers who, in combat, could be forced to remain sealed inside their vehicle for several days.

That misery transfers directly, forcefully away through the air, smacking into and dabbing onto the enemy.

This method of warfare aims to kill the enemy's spirit and psyche. The patent conveys this fact in spirited, if not belletristic, language: 'Except damaging factors, significance of which is secondary in this case, the military psychological positive effect takes place: comprehension of the facts of "delivering" and distribution on enemy territory (on equipment and uniform of an enemy) by the staff as well as the opportunity of informing other soldiers and the enemy about it. As a result, in addition to the basic purpose reaching (full wastes removal) additional military-psychological and military-political effects are achieved.'

Semenov, Aleksandr Georgievich (2009). 'Method of Biowaste Removal from Isolated Dwelling Compartment of Military Facility and Device for Its Implementation', Russian patent no. 2,399,858, 19 January.

MAY WE RECOMMEND

'STOOL SUBSTITUTE TRANSPLANT THERAPY FOR THE ERADICATION OF CLOSTRIDIUM DIFFICILE INFECTION: "REPOOPULATING" THE GUT'

by Elaine O. Petrof, Gregory B. Gloor, Stephen J. Vanner,

Scott J. Weese, David Carater, Michelle C. Daigneault, Eric M. Brown, Kathleen Schroeter and Emma Allen-Vercoe (published in *Microbiome*, 2013)

METHODOLOGY   Open Access

Stool substitute transplant therapy for the eradication of *Clostridium difficile* infection: 'RePOOPulating' the gut

OVERBLOWN BEANS

'People's concerns about excessive flatulence from eating beans may be exaggerated.' That conclusion emerges loud and clear at the end of a study published in the *Nutrition Journal*.

Donna Winham, of Arizona State University, and Andrea Hutchins, of the University of Colorado, call their report 'Perceptions of Flatulence From Bean Consumption Among Adults in 3 Feeding Studies'. 'Many consumers avoid eating beans because they believe legume consumption will cause excessive intestinal gas or flatulence', they explain.

Winham and Hutchins had volunteers eat half a cup of beans daily. Every week everyone answered a questionnaire.

In the first week, fewer than half of the bean eaters reported increases in gas production. Then came a further surprise: 'Seventy percent or more of the participants who experienced flatulence felt that it dissipated by the second or third week of bean consumption.'

Winham and Hutchins suggest that beans owe their unhappy reputation to 'psychological anticipation of flatulence problems'.

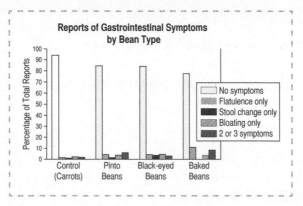

A half cup of beans was consumed daily during the study period

Their opinion of 2011 is opposite, nearly, to one expressed by Geoffrey Wynne-Jones of Waikato Hospital in Hamilton, New Zealand, in 1975. Dr Wynn-Jones published a treatise in *The Lancet*, with the alarming title 'Flatus Retention is the Major Factor in Diverticular Disease'.

Dr Wynne-Jones said: 'Diverticular disease is confined to modern urban communities: flatus retention in a rural, primitive society would be pointless … [The disease] afflicts the cultured, the refined, the considerate … It should be recognised as originating in suppression of a normal bodily function.'

He declared that patients must 'avoid "windy" foods'. He identified beans as a chief example of a 'windy' food.

Winham, Donna M., and Andrea M. Hutchins, 'Perceptions of Flatulence From Bean Consumption Among Adults in 3 Feeding Studies'. *Nutrition Journal* 10 (128): n.p.
Wynne-Jones, Geoffrey (1975). 'Flatus Retention is the Major Factor in Diverticular Disease'. *Lancet* 306 (7927): 211–2.

MAY WE RECOMMEND

'CHILDHOOD CONSTIPATION IS NOT ASSOCIATED WITH CHARACTERISTIC FINGERPRINT PATTERNS'
by C.R. Jackson, B. Anderson and B. Jaffray (published in the *Archives of Disease in Childhood*, 2003)

GOD-AWFUL POLLUTED WATERS

Do the gods pollute? Scientists in India, worried about the public health consequences of immersing idols in lakes and rivers, have been looking anew at water pollution. They hope, and perhaps in some cases pray, to harmonize their medical concerns with some people's religious priorities.

Most of their research has focused on idols of the elephant-headed god Ganesh, created for the annual Ganesh Chaturthi celebration. Once a fairly quiet, mostly private practice, Ganesh Chaturthi now involves large, public festivals in many parts of the country. Researchers have also looked, a little, at the effects of immersing other idols, especially those of the many-armed goddess Durga.

One of the latest studies is called 'Assessment of the Effects of Municipal Sewage, Immersed Idols and Boating on the Heavy Metal and Other Elemental Pollution of Surface Water of the Eutrophic Hussainsagar Lake (Hyderabad, India)'. A team sampled water repeatedly from different parts of the lake, including one spot 'immersed with hundreds of multicolored idols of Lord Ganesh and Goddess Durga', and another near 'the outfall of black-colored, untreated raw sewage containing a collection of industrial effluents'. Sewage, they conclude, accounts for most but not all of the pollution. High levels of zinc, calcium and strontium 'were probably due to the immersed idols painted with multicolors'.

Impact of Ganesh Idol Immersion Activities on the Water Quality of Tapi River, Surat (Gujarat) India

N.C. Ujjania[a] and Azhar A. Multani

Department of Aquatic Biology, Veer Narmad South Gujarat University, Surat (Gujarat) PIN 395007, India

Abstract

In this paper the impact of Ganesh idol immersion on water quality of Tapi River is discussed, for this purpose Ashwanikumar immersion point (Ovara) was selected as sampling station because large number of Ganesh idols immersed on this Ghat of Tapi River. Water samples were collected at morning hours during pre immersion, during immersion and

Some studies concentrate on isolating the effects of idols from those of other sources. 'Impact of Ganesh Idol Immersion

Activities on the Water Quality of Tapi River, Surat (Gujarat) India' tells of sampling the water 'at morning hours during pre-immersion, during immersion and post-immersion periods of Ganesh idols'. The conclusion: the 'main reason of the deterioration of water quality ... is various religious activities', with special blame given to 'the plaster of paris, clothes, iron rods, chemical colours, varnish and paints used for making the idols'.

Several studies examined a lake in the city that suffered India's most famous act of pollution: the 1984 chemical leak from a Union Carbide factory, which resulted in several thousand deaths.

Of particular note, 'Heavy Metal Contamination Cause of Idol Immersion Activities in Urban Lake Bhopal, India', published in 2007, finds that idol immersion has become 'a major source of contamination and sedimentation to the lake water'. It warns that idol-derived heavy metals, especially nickel, lead and mercury, are likely to find their way into 'fishes and birds inhabiting the lake, which finally reach the humans through food'. The authors want to 'educate idol makers' to make their idols small, of non-baked, quick-dissolving clay, and with 'natural colors used in food products'.

Immersion of Ganesha Idols

Immersion site	1998		1999		2000	
	Number	Weight	Number	Weight	Number	Weight
Upper lake	10,076	129.7	7,704	167.9	15,531	390.0
Lower lake	2,528	15.6	2,587	54.9	3,980	92.2

Immersion of Durga Idols

Immersion site	1998		1999		2000	
	Number	Weight	Number	Number	Weight	Number
Upper lake	859	129.3	1,508	125.3	1,301	99.2
Lower lake	543	44.3	45	2.4	95	6.1

Lord Ganesh versus Goddess Durga, as found in Hussainsagar Lake, Hyderabad, India (weight in tonnes)

Reddy, M. Vikram, K. Sagar Babu, V. Balaram and M. Satyanarayanan (2012). 'Assessment of the Effects of Municipal Sewage, Immersed Idols and Boating on the Heavy Metal and Other Elemental Pollution of Surface Water of the Eutrophic Hussainsagar Lake (Hyderabad, India)'. *Environmental Monitoring and Assessment* 184 (4): 2012, 1991–2000.

Ujjania, N.C., and Azhar A. Multani (2011). 'Impact of Ganesh Idol Immersion Activities on the Water Quality of Tapi River, Surat (Gujarat) India'. *Research Journal of Biology* 1 (1): 11–5.

Vyas, Anju, Avinash Bajpai, Neelam Verma and Savita Dixit (2007). 'Heavy Metal Contamination Cause of Idol Immersion Activities in Urban Lake Bhopal, India'. *Journal of Applied Sciences and Environment Management* 11 (4): 37–9.

Bajpai, Avinash, Anju Vyas, Neelam Verma and D.D. Mishra (2008). 'Effect of Idol Immersion on Water Quality of Twin Lakes of Bhopal with Special Reference to Heavy Metals'. *Pollution Research* 27 (3): 517–22.

Vyas, Anju, Avinash Bajpai and Neelam Verma (2008). 'Water Quality Improvement after Shifting of Idol Immersion Site: A case study of Upper Lake, Bhopal, India'. *Environmental Monitoring and Assessment* 145 (1/3): 437–43.

Aniruddhe, Mukerjee (n.d.). 'A Case Study of Idol Immersion in the Context of Urban Lake Management Religious Activities and Management of Water Bodies'. Jabalpur Municipal Corporation, Jabalpur, Madhya Pradesh, India, http://parisaraganapati.net/archives/83.

IN BRIEF

'COW DUNG INGESTION AND INHALATION DEPENDENCE: A CASE REPORT'

by Praveen Khairkar, Prashant Tiple and Govind Bang (published in the *International Journal of Mental Health and Addiction*, 2009)

The authors, at Datta Meghe Institute of Medical Sciences, Wardha, Maharashtra, India, report: 'Although abuse of several unusual inhalants had been documented, addiction to cow dung fumes or their ashes has not been reported in medical literature as yet. We are reporting a case of cow dung dependence in ingestion and inhalational form.'

NAPPY ODOURS

When a mother compares and contrasts the stench from her baby's nappies with that from those of someone else's baby, the question of disgust arises. The question drove a team of psychologists to do an experiment. Richard Stevenson and Trevor Case, of Macquarie University in Sydney, Australia, and Betty Repacholi of the University of Washington in Seattle issued a report called 'My Baby Doesn't Smell As Bad As Yours: The Plasticity of Disgust'. It appeared in a 2006 issue of the journal *Evolution and Human Behavior*.

Stevenson, Case and Repacholi present their work as an addition to a body of earlier smelly investigation. Theirs accords, they say, with a 'line of research on interpersonal preferences for armpit odour'. That old underarm research 'reveals that we have a preference for our own body odours and those from close kin'.

For their experiment, the team crafted a simple procedure: 'mothers of infants were presented with a series of trials in which they smelled concealed samples of their own baby's faeces-soiled diaper and those of someone else's baby.' They tested thirteen mothers, whose productive little ones were aged between six and twenty-four months.

Each mother smelled the contents of a series of buckets. Each bucket housed a soiled diaper, with a special opening that 'prevented participants from seeing the contents of the bucket, while allowing the odour to be sampled.'

The researchers inserted some twists. Sometimes a bucket would be labelled, saying its load came from a particular baby. Sometimes that label identified the wrong baby – thus posing a severe, almost unfair test of the mother's power to discriminate her progeny's output from all others.

The mothers knew their stuff. The Stevenson, Case, Repacholi report concludes that: 'whether the stimuli were correctly labelled, mislabelled, or given no label, mothers rated their own baby's soiled diaper as less disgusting than someone else's baby's diaper.'

The same team, but with Megan Oaten of Macquarie University in place of Repacholi, later examined disgust related to a different aspect of human intimacy: the sights, sounds, textures and odours emanating from step 1 of the baby-manufacturing process. That new report, called 'Effect of Self-Reported Sexual Arousal on Responses to Sex-Related and Non-Sex-Related Disgust Cues', was published in the *Archives of Sexual Behavior*.

Stevenson, Case and Oaten share how (but not why) ninety-nine men agreed to subject themselves to whatever the team would ask them to see, hear, feel or smell. The whatever included images of a 'scar on naked women' and 'pollution'; sounds of someone vomiting and of someone (presumably someone else) performing fellatio; textures comprised of cold pea and ham soup and four lubricated condoms; faecal odours and rotting-fish odours.

A few of the men had first looked at erotic images. Those sexually aroused men 'reported being significantly less disgusted' than the unaroused men by sex-related sights, sounds, feels and smells.

Case, Trevor L., Betty M. Repacholi and Richard J. Stevenson (2006). 'My Baby Doesn't Smell as Bad as Yours: The Plasticity of Disgust'. *Evolution and Human Behavior* 27 (5): 357–65.

Stevenson, Richard J., Trevor L. Case and Megan J. Oaten (2011). 'Effect of Self-Reported Sexual Arousal on Responses to Sex-Related and Non-Sex-Related Disgust Cues'. *Archives of Sexual Behavior* 40 (1): 79–85.

IN BRIEF

'PARENTAL CONSUMPTION OF NESTLING FECES: GOOD FOOD OR SOUND ECONOMICS?'
by Peter L. Hurd, Patrick J. Weatherhead and Susan B. McRae (published in *Behavioral Ecology*, 1991)

CALL FOR INVESTIGATORS

Which are more disgusting to touch – dry powders or slimy gels? A team of investigators from the department of psychology at the University of Miami set up a series of experiments to find out – bearing in mind that, until now, touch-related disgust responses have been largely overlooked in scholarly literature.

Participants touched substances of varying moistures, temperatures and consistencies, rating each on disgustingness. 'Results show that participants rated wet stimuli and stimuli resembling biological consistencies as more disgusting than dry stimuli and stimuli resembling inanimate consistencies, respectively', reported the researchers, led by Debra Lieberman, in 'A Feel for Disgust: Tactile Cues to Pathogen Presence' in the journal *Cognition & Emotion*.

Experimental subjects rated the materials in four categories:

- How disgusting
- How disgusting to put in mouth
- How willing to touch again
- How appealing

After correspondence with the authors, we are able to report the exact recipe for the 'disgusting' material used in this experiment. 'The dough mixture was made using 2 cups of flour, 2 cups of water, 1 cup of salt, and 1 tablespoon of cream of tartar, yielding a consistency similar to Play-Doh. Each dough stimulus was formed into a fist-sized ball.' The non-disgusting control material was cotton rope.

The authors note that new recipes may be required for future experiments: 'Additional studies might select more mucoid, slimy and viscous textures rated as highly disgusting to examine the range of tactile properties associated with the disgust response.' We now ask our readers to share their recipes for this scientific endeavour.

Send your recipe to marca@improbable.com with the subject line:
DISGUST PROTOCOL

THE RAT-CATCHER'S ART

England's professional rat-catching community produced at least two instructive books during the Victorian years.

Studies in the Art of Rat-Catching, by Henry C. Barkley, went on sale in London in 1896. Avowedly educational, it's also a rambling entertainment that finishes up with this jolly sentiment: 'I have heard from half a dozen head-masters of schools that they find the art of rat-catching is so distasteful to their scholars, and so much above their intellect, and so fatiguing an exercise to the youthful mind, that they feel obliged to abandon the study of it and replace it once more by those easier and pleasanter subjects, *Latin* and *Greek*.'

Two years later, Ike Matthews, in Manchester, published his *Full Revelations of a Professional Rat-Catcher After 25 Years' Experience*. It is a more scholarly trove of professional knowledge about rat catchers and about economics.

High standards, Matthews maintains, are essential on the job. 'I maintain that it is a profession, and one that requires much learning and courage. I have found this out when I have been under a warehouse floor, where a lot of Rats were in the traps, and I could not get one man out of 50 to come under the floor and hold the candle for me, not to mention helping me to take the live Rats out of the traps.'

The learned know that some risks are less dire than the public believes: 'a good many people seem to think that if a man puts his hand into a bagful of Rats they will bite him, but I can assure you that a child could do the same thing and not be bitten. Should there be only two or three in the bag, then they will bite, but not in the event of there being a good number.'

One must acquire social skills to handle the occasional awkward moments. The rat catcher 'sometimes experiences difficulties in travelling on the railway', writes Matthews. 'I have often entered an empty third-class carriage, sent my dog under the seat, and put the Rat cage there also. The carriage would fill with passengers, and upon reaching my destination I would take from under the seat my cage full of live Rats, to the amusement of some and the disgust of others.

'I have also entered a railway carriage with my cage of rats when there were passengers in, one or two of whom would generally object to live Rats being in the same compartment.'

The professional, Matthews, explains, 'has always one resource open to him when he has finished a job according to contract (catching say 40 or 50 Rats), should there be a dispute about the price and the people decline to pay the bill, then he has the expedient of letting the Rats at liberty again in the place where he had caught them. Most people will pay the price you send in rather than have the Rats turned loose again.'

Barkley, Henry C. (1896). *Studies in the Art of Rat-Catching*. London: John Murray.
Matthews, Ike (1898). *Full Revelations of a Professional Rat-Catcher After 25 Years' Experience*. Manchester: The Friendly Societies' Printing Company.

MAY WE RECOMMEND

'DETERMINATION OF FAVORITE COMPONENTS OF GARBAGE BY DOGS'
by Bonnie V. Beaver, Margaret Fischer and Charles E. Atkinson (published in *Applied Behaviour Science*, 1992)

The authors, at Texas A&M University and E.I. du Pont de Nemours and Company, Orange, Texas, report: 'This study was to determine which commonly found household garbage odors would be the most attractive to dogs. Thirteen items were tested as fresh odors or were aged for 72 h before testing. Each of the 325 paired odor combinations was tested on 12 dogs. Fresh odors were preferred to aged ones, and meat odors were chosen more often than non-meat odors. Fried liver with onions and baked chicken were the top ranking fresh and old odors. Liver was the highest ranking raw meat.'

AN IMPROBABLE INNOVATION

'ODOR TESTING APPARATUS'
a/k/a cat-box-smell testing chambers, by Henry E. Lowe (US patent no. 4,411,156, granted 1983)

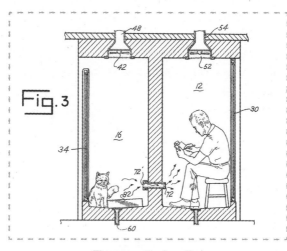

The apparatus in action

THE ART OF UNBUILDING

The engineers who designed and built the very tallest, most skyscraping skyscrapers said very little about whether – let alone how – someone could safely disassemble such a colossus, should the need arise. About the only person who thought about it long and hard was the writer and illustrator David Macauley. Decades ago, Macauley published a children's book called *Unbuilding*. It explains, in words and highly detailed drawings, how to carefully, lovingly take apart the Empire State Building.

In the race to build ever-taller buildings, a problem lurks. These towers are so costly, and the real estate market unpredictable enough, that some of them could become financial failures. To simply abandon something nearly a quarter of a mile tall (the Empire State Building), then let it rot and crumble, would be unneighbourly on a grand scale.

Indeed, how would one disassemble one of the tallest sky-scrapers without risking huge damage to its neighbourhood? And how would this work financially, if the tower needed to be disassembled because the owners ran out of money?

David Macauley tells how, if you have the money, to unbuild in a way that is 'practical and safe'. You take the building down 'floor by floor in the reverse order in which it had been built'.

These days, architecturally, things are looking up. The tallest building, currently the Burj Khalifa in Dubai, at 829.8 metres (2,722 feet) high is more than twice the height of the Empire State Building. In August 2011, the architecture firm Adrian Smith + Gordon Gill announced plans to erect something even taller, the 1,000-metre-tall Kingdom Tower, in Jeddah, Saudi Arabia.

Financially ... well, it's a thrilling time for new skyscraper-owners, and perhaps soon for their neighbours.

The book *Unbuilding*, published in 1980, is a fantasy about the future. In it, the 'Greater Riyadh Institute of Petroleum' needs a new headquarters. A Saudi Arabian prince decides, for symbolic purposes, to purchase the Empire State Building, and move it from New York City to Riyadh. (He chooses, on the basis of cost, to transport only those 'portions of the building necessary to recreate its appearance', and build everything else.)

New Yorkers are at first outraged at the thought of losing their most beloved building. But after the Prince Ali offers the city new parks, museums, and other compensation, nearly everyone agrees to let him take the Empire State Building.

The book at this point has a passage that reads very differently than it did in 1980: 'One desperate but clever preservationist suggested that the twin towers of the World Trade

Center be offered instead – both for the price of the Empire State. In declining the offer Ali suggested that he would be willing to consider pulling them down as a goodwill gesture. With this final show of generosity all remaining resistance crumbled.'

Macauley, David (1980). *Unbuilding.* Boston: Houghton Mifflin.

IN BRIEF

'THE EXPLODING TOILET AND OTHER EMERGENCY ROOM FOLKLORE'
by Robert D. Slay (published in the *Journal of Emergency Medicine,* 1986)

HOW TO REMOVE THE UNMENTIONABLE, DIGITALLY

Nurses, among the most respected persons in our society, must acquire some skills that non-medical people find embarrassing, disgusting and maybe even childish. Such knowledge can be difficult to obtain from the standard medical books and journals. A monograph called 'How to Perform a Digital Removal of Faeces' aims to remedy one such gap in the literature.

Gaye Kyle, a Senior Lecturer at the Faculty of Health and Human Science at Thames Valley University in Slough, researched the subject in depth. She published her findings in the journal *Continence Essentials.*

There exists an official document that purports to provide the digital-removal information a nurse needs to know. But Kyle finds that document wanting. She complains that 'The publication of "Digital Rectal Examination and Manual Removal of Faeces – Guidance for Nurses by the Royal College of Nursing" addressed many issues concerning the professional and legal aspects of the manual (digital) removal of faeces. However, the document did not give detailed guidance on how to actually carry out the procedure.'

Kyle is can do when it comes to how to. What must be discussed, she discusses, prissy diplomatic mincing be damned. 'Digital removal of faeces is a procedure that many healthcare workers are not confident about performing', she writes. 'However, in some patients it is a necessary part of their routine bowel care.'

Some aspects of the situation strike her as dangerous and ludicrous. 'Some nurses are actively refusing to undertake digital removal of faeces on spinal cord injury patients either because they have not been trained or, even more alarmingly, because they think they are not allowed to perform the procedure at all.'

Kyle uses plain language to describe the entire procedure, listing twenty-five distinct, specific actions. These range from the philosophical to the hands-in. For each action, she states a rationale, removing the guesswork that would stymie many a novice or unconfident health care professional.

Why should one ensure privacy? 'To help the patient relax and minimise embarrassment'.

Why should one 'place water-based lubricating gel on gloved index finger … for patients receiving this procedure on a *regular* basis'? The reason is practical and also common-sensical: 'To facilitate easier insertion of index finger'.

Kyle explicates technical minutiae, but only when and where such is needed to provide a clear, unambiguous understanding. 'Gently rotate the finger 6–8 times in a clockwise motion and withdraw', she directs, then goes on to tell how many times the rotation may be repeated, and how many or few minutes should be allowed between each round of stimulation.

Gaye Kyle has given us a case study in the way vital knowledge can remain hidden and difficult to get at, especially when it pertains to matters or matter that can remain hidden and difficult to get at.

Kyle, Gaye (2008). 'How to Perform a Digital Removal of Faeces'. *Continence Essentials* 1: 126–30.

Irwin, Karen (2008). 'Digital Rectal Examination & Manual Removal of Faeces'. NHS Primary Care Trust Bolton, February, http://www.docstoc.com/docs/71006826/Digital-Rectal-Examination-_-Manual-Removal-of-Faeces.

MAY WE RECOMMEND

'COLONIC GAS EXPLOSION: IS A FIRE EXTINGUISHER NECESSARY?'

by J.H. Bond and M.D. Levitt (published in *Gastroenterology*, 1979)

NINE

STUFF IN THE AIR (AND ELSEWHERE)

MAY WE RECOMMEND

'RECTAL IMPALEMENT BY PIRATE SHIP: A CASE REPORT'
by M. Bemelman and E.R. Hammacher (published in *Injury Extra*, 2005)

Some of what's in this chapter: Greetings with potent bugs • Mighty (Marm- and Vege-) mites for mighty few nations • Mozart killer notions • Sudden late pacemaker boom • Too-close look at a pipe • A surprise for the self-auto-shocked snakebit marine • Dragonflies on a black gravestone • U for the ducks • Pop-up pop-off procedure, for bears • The further adventures of Troy Hurtubise • Upside-down sunken-dinosaur scheme • Approaches to parachuting • Insurance for clowns • The tragedy of pizza delivery • Cadaverine and putrescine up your nose • Crimefighting with dolls • Brain extraction on the quick • Dead mules in a literary niche

A HANDY GUIDE TO PATHOGENS

How many pathogens per handshake? Is it dangerous to shake hands at a school graduation?

Dr David Bishai and a team at Johns Hopkins University in Baltimore, Maryland, USA, did a small experiment. They wanted to gauge whether the people most at peril should worry about it.

Their report, 'Quantifying School Officials' Exposure to Bacterial Pathogens at Graduation Ceremonies Using Repeated Observational Measures', was published in the *Journal of School Nursing*. 'This study was designed to measure the degree to which principals and deans are potentially exposed to the risk of pathogen acquisition as part of their occupational duties to shake hands.'

The write-up has some small human touches.

The team recruited officials who had leading roles in graduation ceremonies at elementary, secondary and post-secondary schools in the state of Maryland. Fourteen authority figures agreed to be the subjects of the experiment. Beforehand, each of the fourteen washed with an alcohol-based sanitizer. Then, and afterwards when all the handshaking was done, 'each of the participant's hands was set on a clean drape and swabbed from the base of the thumb to the side of index finger and then around the edges of the other fingers to account for all possible areas for hand contamination during a handshake'.

The risk is pretty small, the results imply. Only two of the fourteen school officials had pathogenic bacteria on hand post-graduation – and only one of those was on the right, shaking hand. Twirling the numbers for perspective, the study explains there is a '0.019% probability of acquiring a pathogen per handshake'.

The researchers point out many reasons why their study is just a preliminary, quick sketch of the story. They examined only a few school officials, and tested for only two kinds of pathogens. Medical science is not clear yet on the prevalence of those pathogens on people's hands in general. Nor is it clear that the microbes' mere presence on the outside (rather than inside) of the body is indicative of danger.

And school graduations are just a sliver of the human experience: 'Graduates may have a different level of infectiousness from other members of the community with whom one might shake hands, making our results less useful to politicians, business executives and clergy.'

Bishai, David, Liang Liu, Stephanie Shiau, Harrison Wang, Cindy Tsai, Margaret Liao, Shivaani Prakash and Tracy Howard (2011). 'Quantifying School Officials' Exposure to Bacterial Pathogens at Graduation Ceremonies Using Repeated Observational Measures'. *Journal of School Nursing* 27 (3): 219–24.

MORE THAN JUST A CONDIMENT

Marmite, the born-in-Britain foodstuff with a powerful taste and a whiff-of-superhero-comic-book name, is more than just a condiment. Marmite, together with its younger, Australian-borne kinsman Vegemite, is an ongoing biomedical experiment.

Streaky dabs of information appear here and there, spread thin, on the pages of medical journals dating back as far as 1931.

The 1930s were a sort of golden period for Marmite. A steady diet of Marmite reports oozed deliciously from several medical journals. Likely many physicians ingested them whilst munching Marmite on toast.

Dr Alexander Goodall of the Royal Informary of Edinburgh regaled readers of *The Lancet* with a case report called 'The Treatment of Pernicious Anæmia by Marmite'. Goodall told how a *British Medical Journal* article, published the previous year, had inspired him and benefited his patients: 'The publication by Lucy Wills of a series of cases of "pernicious anaemia of pregnancy" and "tropical anaemia" successfully treated by marmite raises many questions of importance … Since the publication of Wills's paper I have treated all my "maintenance" cases with marmite. Without exception these have done well.'

Two weeks later, also in *The Lancet*, Stanley Davidson of the University of Aberdeen disagreed. 'It would be very unwise at the present stage', he wrote, 'to suggest that marmite can replace liver and hog's stomach preparations'.

MARMITE IN SPRUE.

To the Editor of THE LANCET.

SIR,—In your issue of April 16th (p. 829) Dr. Janet M. Vaughan and Dr. Donald Hunter write on marmite in the anæmia of idiopathic steatorrhœa, and (on p. 837) is reported a contribution by Dr. Lucy Wills in the discussion at the Section of Tropical Medicine of the Royal Society of Medicine on the megalocytic anæmias of the tropics. May I recall that in the opening paper of a discussion on the treatment of sprue at the same Society on Dec. 3rd, 1923, I mentioned that I had for long used marmite in the treatment of sprue, nearly always complicated by

Lancet readers also got to learn about 'Marmite in Sprue', 'The Treatment by Marmite of Megalocytic Hyperchromic Anaemia: Occurring in Idiopathic Steatorrhœa', and 'The Nature of the Hæmopoietic Factor in Marmite'.

Vegemite starred quietly in a 1948 monograph in the *Journal of Experimental Biology* entitled 'Studies in the Respiration of *Paramecium caudatum*'. Beverley Humphrey and George Humphrey of the University of Sydney described how they grew and nurtured their microbes: 'The culture medium consisted of 5 milliliters of Osterhout solution and 5 milliliters of 20% Vegemite suspension in 1 liter of distilled water. The Vegemite is a yeast concentrate manufactured by the Kraft-Walker Cheese Co. Pty. Ltd., Australia, and served to support a rich bacterial flora upon which the Protozoa fed.' Humphrey and Humphrey's Vegemite adventure contributed, they said, to 'the slow advance of our knowledge of the nutrition of most types of Protozoa'.

A 2003 paper called 'Vegemite as a Marker of National Identity', published in the journal *Gastronomica*, contributed to the slow advance of knowledge of native Australians' liking for the food stuff. Paul Rozin and Michael Siegal surveyed a few hundred students at the University of Queensland, yielding up a table of numbers contrasting their relative (and in many cases relatively large) enjoyment of that substance in comparison with coffee and other common foods. The numbers indicate that Vegemite and carrots enjoyed about equal favour, not as high as chocolate but towering far above sardines or Marmite.

Liking for Vegemite and Other Foods in Native-born Australians

Item	Mean	Correlation with Vegemite
Vegemite	6.52	n/a
chocolate	8.03	-.06
steak	6.62	-.04
peanut butter	6.10	.14
milk	6.87	-.04
carrots	6.60	.09
butter	5.58	.04
apples	6.89	.01
Marmite	2.64	.18
sardines	3.52	-.05
lamb chops	5.76	-.01
broccoli	6.26	.09
black coffee	3.25	.10
pepper	5.95	.05

Table 1 from 'Vegemite as a Marker of National Identity'

In a very few cases, researchers thought they could see hints of a darker side to Vegemite and Marmite.

A 1985 report called 'Vegemite Allergy?', published in the *Medical Journal of Australia*, told of a fifteen-year old girl with asthma: 'She has noted over the last 2–3 years that ingestion of Vegemite, white wine or beer seems to induce wheezing within a short period of time.' The doctors concluded that hers was a 'suspicious theory'.

Four years later, Dr Nigel Higson of Hove issued a bitter warning in the *British Medical Journal* under the headline 'An Allergy to Marmite?'. Higson wrote: 'Some health visitors advise mothers to put Marmite on their nipples to break the child's breast feeding habit; in a susceptible child this action might possibly be fatal.'

Goodall, Alexander (1932). 'The Treatment of Pernicious Aaemia by Marmite'. *The Lancet* 220 (5693): 781–2.

Wills, Lucy (1931). 'Treatment of "Pernicious Anaemia of Pregnancy" and "Tropical Anaemia" with Special Reference to Yeast Extract as a Curative Agent'. *British Medical Journal* i: 1059–64.

Wills, Lucy (1933). 'The Nature of the Haemopoietic Factor in Marmite'. *Lancet* 1: 1283–6

Davidson, Stanley (1932). 'Marmite in Pernicious Anaemia'. *The Lancet* 220 (5695): 919–20.

Rogers, Leonard (1932). 'Marmite in Sprue'. *The Lancet* 219 (5669): 906.

Vaughan, Janet M., and Donald Hunter (1932). 'The Treatment by Marmite of Megalocytic Hyperchromic Anæmia: Occurring in Idiopathic Steatorrhoea (Coeliac Disease)'. *The Lancet* 219 (5668): 829–34.

Humphrey, Beverley A., and George F. Humphrey (1948). 'Studies in the Respiration of *Paramecium caudatum*'. *Journal of Experimental Biology* 25 (June): 123–34.

Rozin, Paul, and Michael Siegal (2003). 'Vegemite as a Marker of National Identity'. *Gastronomica* 3 (4): 63–7.

van Asperen, P.P., and A. Chong (1985). 'Vegemite Allergy?'. *Medical Journal of Australia* 142 (3): 236.

Higson, Nigel (1989). 'An Allergy to Marmite?'. *British Medical Journal* 298 (6667): 190.

MOZART'S DARK DEATH

Wolfgang Amadeus Mozart has died a hundred deaths, more or less. Here's a new one: darkness.

Doctors over the years have resurrected the story of Mozart's death again and again, each time proposing some alternative horrifying medical reason why the eighteenth century's most celebrated and prolific composer keeled over at age thirty-five. One monograph suggests that Mozart died from too little sunlight.

The researchers give us a simple theory. When exposed to sunlight, people's skin naturally produces vitamin D. Mozart, towards the end of his life, was nearly as nocturnal as a vampire, so his skin probably produced very little of the vitamin. (The man failed to take any vitamin D supplements to counteract that deficiency. But that wasn't Mozart's fault. Only much later, in the 1920s, did scientists identify a clear link between vitamin D, sunlight and good health. Vitamin D supplements did not go on sale in Salzburg and Vienna, Mozart's home towns, until many years after that.)

Stefan Pilz (who, if he plays his cards right, will hereafter be known as 'Vitamin' Pilz) and William B. Grant published their report, called 'Vitamin D Deficiency Contributed to Mozart's Death', in the journal *Medical Problems of Performing Artists*. Pilz is a physician/researcher at the Medical University of Graz, Austria. Grant is a California physicist whose background is in optical and laser remote sensing of the atmosphere, and atmospheric sciences.

Pilz and Grant explain: 'Mozart did much of his composing at night, so would have slept during much of the day. At the latitude of Vienna, 48° N, it is impossible to make vitamin D from solar ultraviolet-B irradiance for about six months of the year. Mozart died on 5 December 1791, two to three months into the vitamin D winter.'

They acknowledge the existence of competing medical theories. They do not bother mentioning the possibility, depicted in Peter Shaffer's 1969 play *Amadeus*, that a rival composer did him in. Other academic studies do examine the evidence for poisoning; most conclude that that evidence is lame.

Rival doctors and historians have presented arguments, in medical and other academic journals, that Mozart perished from acute rheumatic fever, bacterial endocarditis, streptococcal septicemia, tuberculosis, cardiovascular disease, brain haemorrhage, hypertensive encephalopathy, congestive heart failure, uremia secondary to chronic kidney disease, pyelonephritis, congenital urinary tract anomaly with obstructive uropathy, bronchopneumonia, haemorrhagic shock, post-streptococcal Henoch-Schönlein purpura, polyarthritis, trichinellosis, amyloidosis and quite a few other unpleasantnesses.

Other studies have tried to tease out biomedical causes for some of Mozart's eccentric behaviour. Two of the more abstruse are by Benjamin Simkin. In 1999 he wrote about a concept called 'Pediatric Autoimmune Neuropsychiatric Disorder Associated with Streptococcal infection' (PANDAS). The study is called 'Was

PANDAS Associated with Mozart's Personality Idiosyncrasies?'. It expanded on Simkin's curse-filled 1992 monograph in the *British Medical Journal* called 'Mozart's Scatological Disorder'.

Distribution of Scatological Letters in the Mozart Family's Correspondence

	Number of letters	Number (%) scatological
Wolfgang Amadeus Mozart	371	39 (10.5%)
Maria Anna Mozart (mother)	40	1 (2.5%)
Nannerl Mozart (sister)	15	1 (6.7%)
Leopold Mozart (father)	319	1 (0.3%)

From 'Mozart's Scatological Disorder' (1992). The author goes on to inventory the scatological terms in Mozart's letters, including 'arse', 'muck', 'piddle or piss' and 'fart'.

Grant, William B., and Stefan Pilz (2011). 'Vitamin D Deficiency Contributed to Mozart's Death'. *Medical Problems of Performing Artists* 26 (2): 117.

Dawson, William J. (2010). 'Wolfgang Amadeus Mozart – Controversies Regarding His Illnesses and Death: A Bibliographic Review'. *Medical Problems of Performing Artists* 25 (2): 49.

Zegers, Richard H.C., Andreas Weigl and Andrew Steptoe (2009). 'The Death of Wolfgang Amadeus Mozart: An Epidemiologic Perspective'. *Annals of Internal Medicine* 151 (4): 274–8.

Simkin, Benjamin (1999). 'Was PANDAS Associated with Mozart's Personality Idiosyncrasies?'. *Medical Problems of Performing Artists* 14 (3): 113.

— (1992). 'Mozart's Scatological Disorder'. *British Medical Journal* 305 (19 Dec): 1563.

RESEARCH SPOTLIGHT

'SUNSHINE AND SUICIDE INCIDENCE'

by Martin Voracek and Maryanne L. Fisher (published in *Epidemiology*, 2002)

'SOLAR ECLIPSE AND SUICIDE'

by Martin Voracek, Maryanne L. Fisher and Gernot Sonneck (published in the *American Journal of Psychiatry*, 2002)

THE PROBLEM OF EXPLODING PACEMAKERS

People muse that they will, come the day, 'go out with a bang'. A little more often than you might expect, someone or other does exactly that, which is why there came into existence a study called 'Pacemaker Explosions in Crematoria: Problems and Possible Solutions'. Christopher Gale, of St James's University Hospital in Leeds, and Graham Mulley, of the General Infirmary at Leeds, published the report in 2002 in the *Journal of the Royal Society of Medicine*.

The first crematorium pacemaker explosion on record happened in 1976 in Solihull, in the West Midlands. As pacemakers and cremations both became popular, Gale and Mulley explain, after-death explosions came to be expected.

Frequency of Pacemaker Explosions in Crematoria in the UK as Estimated by Crematoria Staff

Frequency of explosions occurring in the UK	Crematoria staff reporting event at this frequency (%)
Never	41
Once every 10 years	27
Once every 5 years	14
Once every 2 years	6
Once a year	3
Greater than once a year	2
Not answered	7

Table 1, showing responses from 188 crematoria
staff who responded to the questionnaire.

In the wake of the Solihull surprise, the *British Medical Journal* published an essay entitled 'Hidden Hazards of Cremation'. It speaks of the incident, and speculates about even worse things. There are individuals, it points out, who, as a result of medical procedures, contain tiny amounts of radioactive substances:

yttrium-90, iodine-131, gold-198, phosphorus-32 and their ilk. 'There is a possibility', the anonymous author confides, 'that an explosion (or some other event) during the cremation of a radioactive corpse could produce a blow-back releasing radioactive smoke or fumes into the crematorium. This risk seems to be largely theoretical, but ...'

Soon, two questions were added to Form B of the government-mandated Cremation Act Certificate, which a physician fills out prior to a cremation. 'Has a pacemaker or any radioactive material been inserted in the deceased (yes or no)?; (b) If so, has it been removed (yes or no)?' (Form B, after being retooled, eventually came to the end of its own life, and was replaced in 2008.)

Gale and Mulley surveyed the managers of all 241 cremation facilities in the UK, asking: '(1) Have you ever had personal experience of pacemaker explosions in crematoria? (2) What do you estimate is the frequency of pacemaker explosions in crematoria?' About half the respondents admitted to 'personal experience' of pacemaker explosions. Gale and Mulley suspected that the actual numbers were more dire. 'Staff may not wish to mention these events', they write, 'and their recall may not be accurate.' Gale and Mulley also concluded that 'crematoria staff rely on accurate and complete cremation forms' – but that the information-gathering/reporting process could, and often did, go astray.

The pair published another report, two years later, to show that pacemakers themselves sometimes go astray. Called 'A Migrating Pacemaker', it tells how Dr Gale failed, using his medically trained hands and fingers, to find the pacemaker in a seventy-nine-year-old deceased man. The man's medical records said there should be one.

Gale got a hand-held metal detector, which showed that the pacemaker had moved elsewhere in the body. He concluded that medical devices don't always stay exactly where surgeons originally placed them, and he recommends using hand-held metal detectors to 'help prevent explosions in crematoria'.

Gale, Christopher P., and Graham P. Mulley (2002). 'Pacemaker Explosions in Crematoria: Problems and Possible Solutions'. *Journal of the Royal Society of Medicine* 95 (7): 353–5.
— (2005). 'A Migrating Pacemaker'. *Postgraduate Medical Journal* 81: 198–9.
n.a. (1977). 'Hidden Hazards of Cremation'. *British Medical Journal* 2 (24 Dec): 1620–1.

IN BRIEF

'AIRBAG DEPLOYMENT AND EYE PERFORATION BY A TOBACCO PIPE'
by F.H. Walz, M. Mackay and B. Gloor (published in the *Journal of Trauma*, 1995)

THE RATTLING TALE OF PATIENT X

The self-inflicted snake/electroshock saga of Patient X turned out, twenty-one years after it entered the public record, to have an earlier chapter that had not made it into print.

That's Patient X, the former US marine who suffered a bite from his pet rattlesnake. Patient X, the man who immediately after the bite insisted that a neighbour attach car spark plug wires to his lip, and that the neighbour rev up the car engine to 3,000 rpm, repeatedly, for about five minutes. Patient X, the bloated, blackened, corpse-like individual, who subsequently was helicoptered to a hospital where Dr Richard C. Dart and Dr Richard A. Gustafson saved his life and took photographs of him.

Patient X, who featured in the treatise called 'Failure of Electric Shock Treatment for Rattlesnake Envenomation', which Dart and Gustafson published in the *Annals of Emergency Medicine* in 1991, in consequence of which the three men shared the Ig Nobel Prize in medicine in 1994.

Though rattlesnake bites can be deadly, there is a standard, reliable treatment – injection with a substance called 'antivenin'. Patient X preferred an alternative treatment. The medical report explains: 'Based on their understanding of an article in an outdoorsman's magazine, the patient and his neighbour had previously established a plan to use electric shock treatment if either was envenomated.'

One day while Patient X was playing with his snake, the serpent embedded its fangs into Patient X's upper lip. The neighbour sprang into action. As per their agreement, he laid Patient X on the ground next to the car, and affixed a spark plug wire to the stricken fellow's lip with a small metal clip. The rest you know, at least in outline.

Gustafson eventually went to work for a US Defense Department official agency involved in 'countering weapons of mass destruction'. In 2012 he told me about a conversation he had with Patient X, but did not mention in the medical report: 'I started off by telling Patient X, who was a US marine, that I had been an active duty Navy corpsman. Therefore, I was asking Patient X about his medical history and got around to asking him if he had ever been envenomed by a snake before and if so what treatment did he receive. He said that he had been "bitten" a couple years ago on Okinawa. I then asked him if it had been a Habu, a type of cobra common to Okinawa, and he said yes. I then asked if he had been bitten on his middle finger of his left hand, up in the northern training area, and helicoptered down to Lester Naval hospital to the ER [emergency room]. He said hesitantly "yes". I then informed Patient X that I was the corpsman who was assigned to his care, to monitor his condition for the six hours he was in the ER, prior to being admitted to the hospital as an in-patient.'

Dart, Richard, and Richard Gustafson (1991). 'Failure of Electric Shock Treatment for Rattlesnake Envenomation'. *Annals of Emergency Medicine* 20 (6): 659–61.

RESEARCH SPOTLIGHT

'THE RATIO OF 2ND TO 4TH DIGIT LENGTH: A NEW PREDICTOR OF DISEASE PREDISPOSITION?'

by John T. Manning and Peter E. Bundred (published in *Medical Hypotheses*, 2000)

Professor Manning hypothesizes: 'The ratio between the length of the 2nd and 4th digits is: (a) fixed in utero; (b) lower in men than in women; (c) negatively related to testosterone and sperm counts; and (d) positively related to oestrogen concentrations. Prenatal levels of testosterone and oestrogen have been implicated in infertility, autism, dyslexia, migraine, stammering, immune dysfunction, myocardial infarction and breast cancer. We suggest that 2D:4D ratio is predictive of these diseases.'

SIGNALLING A GRAVE ATTRACTION

Ecological traps for dragonflies in a cemetery: the attraction of *Sympetrum* species (Odonata: Libellulidae) by horizontally polarizing black gravestones

GÁBOR HORVÁTH*, PÉTER MALIK*, GYÖRGY KRISKA† AND HANSRUEDI WILDERMUTH‡
*Biooptics Laboratory, Department of Biological Physics, Physical Institute, Eötvös University, Budapest, Hungary
†Group for Methodology in Biology Teaching, Biological Institute, Eötvös University, Hungary
‡Institute of Zoology, University of Zürich, Zürich, Switzerland

SUMMARY

1. We observed that the dragonfly species *Sympetrum flaveolum, S. striolatum, S. sanguineum, S. meridionale* and *S. danae* were attracted by polished black gravestones in a Hungarian cemetery.

2. The insects showed the same behaviour as at water: (i) they perched persistently in the immediate vicinity of the chosen gravestones and defended their perch against other

Gábor Horváth, head of the Environmental Optics Laboratory at Eotvos University in Budapest, Hungary, solves mysteries about light and about living creatures. Among these, he discovered that white horses attract fewer flies. He and five colleagues wrote a study called 'An Unexpected Advantage of Whiteness in Horses: The Most Horsefly-proof Horse Has a Depolarizing White Coat', which they published in the *Proceedings of the Royal Society B*.

The researchers experimented with a small number of sticky horses and a large number of horseflies (of the variety called tabanids). The horses were sticky because the scientists had coated them with 'a transparent, odourless and colourless insect monitoring glue [called] Babolna Bio mouse trap'.

The scientists brought those horses – one black, one brown, one white – to a grassy field in the town of Szokolya, Hungary. Every other day, they collected and counted the flies that had become attached to the sticky horses.

The results, tallied over fifty-four summer days: the sticky brown horse Babolna-Bio-mouse-trapped fifteen times as many flies as the sticky white horse. And the black horse, poor thing, trapped a whopping, buzzing twenty-five times as many flies as the white one.

The differences, say the scientists, come from the way light bounces off horsehair. Polarized light – light that's all vibrating in the same direction – attracts horseflies. When that light reflects off dark fur, it stays polarized. But when polarized light glances off white fur, it becomes less polarized, which, to a horsefly, is not so attractive.

At the very end of the report, Horváth alludes coquettishly to another of his experiments, one he has not yet written up. It's a case of animal and animal on animal: 'When white cattle egrets, for example, are sitting on the back of dark-coated cattle and pecking blood-sucking tabanids away from the cattle, tabanids blending in colour with that of their hosts' fur might derive greater protection owing to camouflage.'

Indeed, Gábor Horváth has tackled other colourful questions. He and several collaborators reported that red cars attract insects. Details are in the report 'Why Do Red and Dark-coloured Cars Lure Aquatic Insects?: The Attraction of Water Insects to Car Paintwork Explained by Reflection-polarization Signals', also published in the *Proceedings of the Royal Society B*.

Horváth's team likewise discovered that black gravestones

attract dragonflies. Read about that in a paper called 'Ecological Traps for Dragonflies in a Cemetery: The Attraction of *Sympetrum* species (Odonata: Libellulidae) by Horizontally Polarizing Black Grave-stones', published in the journal *Freshwater Biology*.

Not all Gábor Horváths are equally colourful in their scientific writings. Across town, at Budapest University of Technology and Economics (BUTE), associate professor of electrical engineering Gábor Horváth can boast an impressive array of published research work. But to non-specialists most of it looks comparatively unspectacular – the one slight exception being his monograph entitled 'A Sparse Robust Model for a Linz-Donawitz Steel Converter'.

Horváth, Gábor, Miklós Blahó, György Kriska, Ramón Hegedüs, Baláz Gerics, Róbert Farkas and Susanne Åkesson (2010). 'An Unexpected Advantage of Whiteness in Horses: The Most Horsefly-proof Horse Has a Depolarizing White Coat'. *Proceedings of the Royal Society B* 277 (1688): 1643–50.

Kriska, György, Zoltán Csabai, Pál Boda, Péter Malik and Gábor Horváth (2006). 'Why Do Red and Dark-coloured Cars Lure Aquatic Insects?: The Attraction of Water Insects to Car Paintwork Explained by Reflection-polarization Signals'. *Proceedings of the Royal Society B* 273 (1594): 1667–71.

Horváth, Gábor, Péter Malik, György Kriska and Hansruedi Wildermuth (2007). 'Ecological Traps for Dragonflies in a Cemetery: The Attraction of *Sympetrum* species (Odonata: Libellulidae) by Horizontally Polarizing Black Grave-stones'. *Freshwater Biology* 52 (9): 1700–9.

Valyon, József, and Gábor Horváth (2008). 'A Sparse Robust Model for a Linz-Donawitz Steel Converter'. IEEE Transacations on Instrumentation and Measurement 58 (8): 2611–7.

URANIUM IS FOR THE BIRDS

Depleted uranium should, perhaps, be the ammunition of choice for duck hunters. That's the conclusion of a study called 'Response of American Black Ducks to Dietary Uranium: A Proposed Substitute for Lead Shot'.

Published in 1983 in the *Journal of Wildlife Management*, the recommendation has not been much disputed. The study's authors, biologists Susan Haseltine and Louis Sileo, were based at the Patuxent Wildlife Research Center in Laurel, Maryland.

Lead shot is dangerous for ducks, especially if it hits them. When it doesn't hit a duck (or hit another hunter, as sometimes happens), the shot falls into the wetlands. The lead leaches into the muck, slowly poisoning whichever ducks have managed to avoid being shot.

In many hunting areas, lead shot is *verboten*. At the time of the study, steel was being touted as the best alternative to lead. But Haseltine and Sileo pointed out its drawbacks. They wrote that 'Steel shot shells are more expensive than lead shot shells when purchased in a retail outlet; they cannot be used in all guns and have not been well received by some hunters who question their performance on ducks and geese.'

Haseltine and Sileo credit the idea – substituting uranium for steel – to metallurgist Carl A. Zapffe of Baltimore. Zapffe was no slouch about steel; witness his 1948 study: 'Evaluation of Pickling Inhibitors from the Standpoint of Hydrogen Embrittlement: Acid Pickling of Stainless Steel'. Zapffe also wrote a book disputing Einstein's theory of special relativity, but that is a separate matter.

Haseltine and Sileo listed what they call the 'attractive characteristics' of depleted uranium as a raw material for making birdshot. 'In its pure form', they wrote, 'it is denser than lead and, in alloys, might be made to produce shot patterns and velocities attractive to hunters and within the effective range for waterfowl. Depleted uranium can be alloyed with many other metals and its softness and corrosiveness can be altered over a wide range.'

But nothing is perfect. 'Negative aspects for potential uranium shot include pyrophoricity [proneness to spontaneously burst into flames] problems with pure depleted uranium, which can be altered by alloying, and the expense of separating depleted uranium from other nuclear waste products.'

Their main argument was that uranium may not be very poisonous even to a duck that, of its own accord, swallows some in pellet form. That is what Haseltine and Sileo sought to verify.

They fed forty ducks a diet of commercial duck mash salted with powdered depleted uranium. None of the ducks died of it, or got sick, or even lost weight. Moreover, the researchers reported, the ducks 'were in fair to excellent flesh' when slaughtered. And so they enthused that 'further examination of this metal as a substitute for lead in shot is justified.' However, no one has yet followed up on this in a big way for hunting anything other than people.

Haseltine, Susan D., and Louis Sileo (1983). 'Response of American Black Ducks to Dietary Uranium: A Proposed Substitute for Lead Shot'. *Journal of Wildlife Management* 47 (4): 1124–9.

Zapffe, Carl A., and M. Eleanor Haslem (1948). 'Evaluation of Pickling Inhibitors from the Standpoint of Hydrogen Embrittlement: Acid Pickling of Stainless Steel', *Wire and Wire Products* 23 (10): 933–9.

Ricker III, Harry H. (2006). 'Dr. Carl Andrew Zapffe: A "Cod" Proposes a "Flying Interferometer"'. *General Science Journal*, 15 November, http://www.gsjournal.net/old/science/ricker24.pdf.

— (2007). 'An Introduction to Dr. Carl A. Zapffe's Classic Paper: A Reminder on E=mc2'. *General Science Journal*, 3 March, http://gsjournal.net/Science-Journals/Research%20Papers-Relativity%20Theory/Download/866.

Zapffe, Carl A. (1982). *A Reminder on E=mc², m=m$_0$(1-v²/c²)$^{-1/2}$, & N=N$_0$e$^{-t/8t}$*. n.a.: privately published.

—. (1985). 'Exodus of Einstein's Special Theory in Seven Simple Steps'. *The Toth-Maatian Review* 3 (4): 1531–5.

MAY WE RECOMMEND

'CARBON MONOXIDE: TO BOLDLY GO WHERE NO HAS GONE BEFORE'
by Stefan W. Ryter, Danielle Morse and Augustine M.K. Choi
(published in *Science STKE*, 2004)

A BEAR TO CROSS

When a big bear approaches, some people choose to quietly stroll away. To give them an extra measure of safety, Anthony Victor Saunders and Adam Warwick Bell invented what they call a 'Pop-up Device for Deterring an Attacking Animal such as a Bear'.

Saunders, a London-based mountain climber, and Bell, a California patent attorney, applied for a patent in 2003, but later abandoned it. They would equip hikers with, essentially, an inflatable doll 'meant to scare away an attacking or aggressive animal such as a bear'. The frightful balloon could also be used against 'elk, moose, mountain lions, buffalo, hippopotamus, rhino, elephant, boar'. They explain that it 'works on the principle of maximizing the apparent size and ferocity of the human, intimidating the bear'.

Events leading to the use of the pop-up device

In the patent application, Saunders and Bell refined their thoughts. Here's how they decided the device must deploy quickly: 'The figure should be fully inflated within less than one minute, or within less than 30 seconds, or preferably within less than 10 seconds, or most preferably five seconds.' The device can be 'incorporated into clothing or luggage [or] into the hilt of a walking-stick' and activated 'by pulling a cord. The figure would inflate and pop up out of the back-pack, presenting the attacking bear with a huge and frightening opponent'.

The bear then gets an escalating series of surprises, beginning with 'one or more explosive "bangs", a fog-horn, or a loud roaring or screaming sound'. The noise is augmented with smells. 'The musky odour of a bear helps convince the attacking bear that he is being faced with a powerful, aggressive and musky opponent.' Then would come 'an odorous or noxious gas or liquid'.

There's also smoke: 'from a typical "smoke-bomb" type of device'.

Some bears are not easily deterred. So 'the deployment of the device may be accompanied by the launching of projectiles. [This] would further confuse, scare and disorientate the bear. Such projectiles could be launched from a mortar or mortar-type device'.

The whole thing, they say, is 'detachable and may be left between human and bear as the human retreats'.

Inventors Saunders and Bell are not the only pioneering individuals to consider novel methods for greeting bears. In this era of 'big science', there are still people who pursue thoughtful, original research, unencumbered by official scientific credentials, academic bureaucracies or government funding. Troy Hurtubise is a fine example of the breed.

At age twenty, out alone panning for gold in the Canadian wilderness, Hurtubise had an encounter of some sort with a grizzly bear. He has devoted the rest of his life to creating a grizzly-bear-proof suit of armour in which he could safely go and commune with that bear. The suit's basic design was influenced by the powerful humanoid-policeman-robot-from-the-future title character in *Robocop*, a film Hurtubise happened to see shortly before he began his intensive research and development work.

Hurtubise is a pure example of the lone inventor, in the tradition of James Watt, Thomas Edison and Nikola Tesla. Regarded by some as a half-genius, by others as a half-crackpot, he has unsurpassed persistence and imagination. He also is very careful. The proof that he is very careful is that he is still alive.

A grizzly bear is tremendously, ferociously powerful. Hurtubise realized that he would be wise to test his suit under controlled conditions prior to giving it the ultimate test. He spent seven years, and by his estimate $150,000 Canadian, subjecting the suit to every large, sudden force he could devise. Volunteers have pushed him off cliffs, rammed him with trucks travelling forty kilometres per hour, and assaulted him with logs, arrows and pickaxes. For almost all of the testing, Hurtubise was locked inside the bulky suit, despite being severely claustrophobic.

The suit is a technical wonder, especially when one realizes that Hurtubise had to assemble it mostly from scrounged materials. Among the components:

- titanium plates
- a fireproof rubber exterior
- joints made of chain mail
- a Tek plastic inner shell
- an inner layer of air bags
- nearly a mile of duct tape

For conceiving, building and personally testing the suit, and for keeping it and himself intact the whole while, Troy Hurtubise was awarded the 1998 Ig Nobel Prize in the field of safety engineering.

Hurtubise has continued to do advanced R&D work. He has had unexpected adventures involving, among other things, NASA, the National Hockey League, an invention to separate oil from sand, a tapped phone, a mysterious nocturnal break-in, getting kicked in the crotch on television by comedian Roseanne Barr, a visit from Al Qaeda hijackers and an encounter in a locked room with two Kodiak bears.

He also created a book in which he tells secrets about the most advanced version of his suit, the fruit of fifteen mostly unfunded years of fevered research and development. Here's a passage that brings together some of the main themes:

> Electronically speaking, the M-7 was right out of a movie. It sported an onboard viewing screen, an onboard computer built into the thigh cavity, a bite-bar on the right forearm, a five-way voice-activated radio system and an electronic temperature monitor. For protection against the grizzly's claws and teeth, the M-7 boasted an entire exoskeleton made up of my newly developed … blunt trauma foam to dissipate the bear's deadly power. Testing on the M-7 [was] short and sweet. A 30-ton front-end loader in fourth gear smashed me through a non-mortared brick wall and I suffered not a bruise. The world watched the test on CNN, and then came the sheer stupidity that nearly cost me my life, the fire test. My bear research suits were never designed for fire.

Bear Man: The Troy Hurtubise Saga is a magnum opus, the tersely told summary of the man's yearnings, frustrations, triumphs and philosophy. The book includes many of Hurtubise's previous writings on these subjects, augmented with a powerful-as-a-riled-up-grizzly collection of previously private photos, philosophy, intellectualizing and emoting.

He shares with us a letter from Her Majesty the Queen, to whom he had sent some lightly fictionalized writings about his personal knowledge of angels. 'This great lady of ladies found the time to read my novellas and to respond to me in a letter through her Lady in waiting', he writes. 'I was so overwhelmed by Her Majesty's kindness that I dedicated the third novella from the series, *The Canadians*, in her honour … As for her son, Prince Charles, his letter to me was stamped confidential.'

Bear Man: The Troy Hurtubise Saga makes a lovely gift for any young girl or boy who might some day have to decide unexpectedly whether to devote a lifetime to inventing, testing and informing the world about new ways to protect themselves against grizzly bears while doing no harm to the animals, all the while struggling to lead a good life and set a fine example for the youth both of today and of the future.

Hurtubise's basic bear-suit research, which brought him the fame and respect he now enjoys, is best seen in the documentary *Project Grizzly*, produced by the National Film Board of Canada in 1996. You can watch it online at www.nfb.ca/film/project_grizzly.

Bell, Adam Warwick, and Anthony Victor Saunders (2005). 'Pop-up Device for Deterring an Attacking Animal such as a Bear'. US patent no. 2005/0028720, 10 February.
Hurtubise, Troy (2011). *Bear Man: The Troy Hurtubise Saga*. Westbrook, ME: Raven House Publishing.

THE POST-MORTEM DINOSAUR KERBLAM

Seagoing dinosaurs did not explode nearly as often as scientists believed, according to a study called 'Float, Explode or Sink: Postmortem Fate of Lung-breathing Marine Vertebrates'. The authors, an all-star team of palaeontologists, pathologists and forensic anthropologists from six institutions in Switzerland and Germany, deflated a hypothesis that had for years lain basking in the intellectual shade. They were addressing the underlying question: why are some dino skeletons scattered across an expanse of sea floor, while others remain fairly intact?

The current adventure started with the discovery of an ichthyosaur skeleton, embedded in rock, in northern Switzerland. This skeleton was oriented weirdly, compared with most such fossils: aligned vertically, with its head down, its feet up.

Someone hypothesized that 'an explosive release of sewer gas' had 'propelled the skull into the sediment'. The subsequent research, resulting in the 'Float, Explode …' paper, tried to figure out whether that was at all likely.

In so doing, the scientists confronted an idea proposed in 1976 by a palaeontologist named Keller. Keller, noting that beached whales fester in sunlight until putrefaction gases bloat and finally burst them, suggested that sunken sea animal carcasses also gassify and go *kerblam*. The Swiss/German study summarizes Keller's idea as follows: 'It was assumed that carcasses which lie on the sea-floor might have exploded or internal organs and bones erupted, and that in so doing, bones as well as foetuses were ejected and ribs were fractured.'

The team scoured reams of research about what happens to dead dolphins, porpoises, whales, seals, turtles and other sea animals. They say that unless such bodies get stranded on a beach, there's little evidence and little reason to expect that they explode.

The team presented an early version of their debunkment in 2004 in St Petersburg, Russia, at the Fifth Congress of the Baltic Medico-Legal Association. They called their lecture 'Did the Ichthyosaurs Explode? A Forensic-medical Contribution to the Taphonomy of Ichthyosaurs'.

Taphonomy, a word that misleadingly suggests both telephones and tap-dancing, is in fact the study of how living things rot and decay. TV crime-scene forensics series present taphonomic adventures week after week, teasing out the likely when, where and how of one or another winsome corpse. This is better. Real-life scientists – Achim Reisdorf, Roman Bux, Daniel Wyler, Mark Benecke and colleagues – had the opportunity to fawn over a corpse way more glamorous than the TV crime drama standard: a sea-monstrous dinosaur.

This is their own take on what actually happened in the mysterious case of the vertical victim: 'The ichthyosaur sank headfirst into the seafloor because of its centre of gravity, as anatomically similar, comparably preserved specimens suggest. The skull penetrated into the soupy to soft substrate until the fins touched the seafloor.'

A few people disagree. *Creation* magazine made a video explaining that no scientist can explain the existence of that upside-down dinosaur – that it deals 'a lethal body blow' to the theory of evolution. Behold their creative reasoning at http://creation.com/creation-magazine-live-episode-53.

Reisdorf, Achim G., Roman Bux, Daniel Wyler, Mark Benecke, Christian Klug, Michael W. Maisch, Peter Fornaro and Andreas Wetzel (2012). 'Float, Explode or Sink: Post-mortem Fate of Lung-breathing Marine Vertebrates'. *Palaeobiodiversity and Palaeoenvironments* 92 (1): 67–81.

Keller, T. (1976). 'Magen- und Darminhalte von Ichthyosauriern des Süddeutschen Posidonienschiefers'. *Neues Jahrbuch für Geologie und Paläontologie Monatshefte* 5: 266–83.

Bux, R., Reisdorf, A., and Ramsthaler, F. (2004). 'Did the Ichthyosaurs Explode? A Forensic-medical Contribution to the Taphonomy of Ichthyosaurs'. Proceedings of the Fifth Congress of the Baltic Medico-Legal Association, St Petersburg, Russia, 6–9 October, p. 69.

Wetzel, Andreas, and Achim G. Reisdorf (2007). 'Ichnofabrics Elucidate the Accumulation History of a Condensed Interval Containing a Vertically Emplaced Ichthyosaur Skull'. In Bromley, Richard Granville (ed.). *Sediment–Organism Interactions: A Multifaceted Ichnology*. Tulsa, Ok.: SEPM (Society for Sedimentary Geology), special publication no. 88, pp. 241–51.

AN IMPROBABLE INNOVATION

'A SAFETY DEVICE FOR USE IN MAKING A LANDING FROM AN AEROPLANE OR OTHER VEHICLE OF THE AIR'

a/k/a a parachuteless emergency glider, by Hermann W. Williams (US patent no. 1,799,664, granted 1931)

MAY WE RECOMMEND

'PARACHUTE USE TO PREVENT DEATH AND MAJOR TRAUMA RELATED TO GRAVITATIONAL CHALLENGE: SYSTEMATIC REVIEW OF RANDOMISED CONTROLLED TRIALS'.

by G.C.S. Smith and J.P. Pell (published in *British Medical Journal,* 2003)

'PARACHUTING FOR CHARITY: IS IT WORTH THE MONEY? A 5-YEAR AUDIT OF PARACHUTE INJURIES IN TAYSIDE AND THE COST TO THE NHS'

by C.T. Lee, P. Williams, and W.A. Hadden (published in *Injury,* 1999)

SEND IN THE CLOWN INSURERS

Clown insurance is for clowns, not for persons potentially afflicted by them. Insurance companies offer it to clowns because clowns – no matter what you may thoughtlessly think of them – are people, and bad things can happen to anyone. Consider this story, shared by 'Posadaclown' on a popular online clowning forum in 2005:

> Just had the worst night of my clowning career last night. Took it up seriously a couple of years ago, joined a local troupe who would perform around my locality (Northumberland), and things were going quite well … inbetween the goat shearing and terrier racing, when I managed to spin a few plates, do a bit of slapstick incorporating a farmers daughter and managed to get some great laughs for a comedy routine where I made a rabbit 'disappear' before running down my trouser leg and away! Things were going really well until last night when I did a performance at a local club. My family were all there, and my gorgeous new wife, who herself likes to don the red nose and baggy trousers from time to time. Anyway, I did a stunt which involved jumping off a trampoline onto a skateboard, I was supposed to shoot offstage but ended up in the front row, where I managed to land right on top of the mayor's wife, breaking her leg. The poor woman was in agony and a few in the crowd took exception and beat me quite badly.

Clown insurance exists, as a distinct product category, thanks to the mathematical discipline called risk assessment. Industry researchers calculate that every year enough bad things happen to enough clowns to reliably yield a profit. Clowns, as a group, perform a benefit/cost calculation; that's why, year after year, they spend money to defend against life's practical jokes.

The UK boasts several suppliers of insurance for clowns. Blackfriars Insurance Brokers (www.blackfriarsgroup.co.uk/business/insuranceforclowns.html), for one, offers public liability clown insurance of up to £5 million cover. Their website boasts, not unkindly, of 'products to meet the business and personal insurance needs of clowns'. Blackfriars also offers insurance for those who hire clowns: 'Clowns employers liability insurance protects the policyholder in respect of your legal liability for personal injury or illness suffered by employees during the course of their employment'.

Foreign clowns, too, can buy clown insurance. Pretty much any clown, anywhere, can join the World Clown Association (worldclown.com). The association, based in the US, offers its members liability insurance 'with coverage of $1,000,000 per occurrence/$2,000,000 aggregate per event'.

Their insurance application form specifically excludes many of the activities that, one can infer, have proved troublesome.

They will not insure a clown for clowning that involves hypnosis, bouncy castles, hot air balloons, sky diving or competition racing. The application form does not distinguish between the numerous forms of racing – foot, horse, camel, bicycle, ski, cigarette boat, dragster, Spitfire, what-have-you.

Other things too, are *verboten* for the clown who wants to be protected by the association's standard clown insurance.

No throwing objects in any manner other than juggling them.

No working with animals, other than 'performing dogs, doves and rabbits'.

No 'pyrotechnics, explosives, fireworks or similar materials'. But the association is not a killjoy. It explicitly makes an exception for 'concussion effects', 'flashpots' and 'smokepots'.

No copyright infringement (that's what they call it, though

they may mean trademark infringement). The association specifically mentions Bugs Bunny and Mickey Mouse as examples. In any event, one would do well to seek professional advice before simultaneously wearing a Mickey Mouse costume and a clown suit.

LIABILITY INSURANCE (Optional) Cost: $139/year per person. Effective May 1 thru April 30
You must be a member of the World Clown Association to purchase the insurance.
☐ You must check box, sign this form and fill in Type of performance to obtain insurance.

Coverage: This is a Comprehensive General Liability Policy provided by an A-rated insurance company with coverage of $1,000,000 per occurrence/$2,000,000 aggregate per event. Deductible: None The price remains the same regardless of number of months covered (The fee is NOT prorated) and includes Premium, Brokerage fee and Association handling fee.

Policy Exclusions: This program is designed for Clowns and Magicians. The following types of performances are NOT ACCEPTABLE.

1. Hypnosis	5. Animals except performing dogs, doves and rabbits
2. Hot Air Balloon Events, Circuses (production), overnight camping Tractor pulls, Rodeo and roping events, motorized events, Mechanical amusement devices, inflatables, rock climb,Bounce Houses, laser tag, sky diving, competition racing.	6. Musicians and Disc Jockeys, except clowns/magicians who engage in these activities as part of their act
3. Pyrotechnics, Explosives, Fireworks or similar materials except "concussion effects, "flashpots", and "smokepots" (Flashboxes covered up to $5,000)	7. Copyright infringement exclusion (i.e. Barney, Mickey Mouse, Bugs Bunny)
4. Production Management or Promotions Management for hire	8. Throwing objects (juggling is acceptable)

The US is blessed with a large population of clowns. American clowns, unlike their counterparts in much of the world, are blessed with a wide variety of vendors eager to sell off-the-shelf clown insurance. American insurers, in agreement with their British brethren, view clowns as an increasingly specialized species of customer.

Not long ago, hypnotists, fire-eaters and people who do face-painting of children at public events were welcome to purchase standard clown-insurance policies. Then came legal climate change. Now everyone, no matter how clownish they believe themselves to be or not to be, should check the details before binding themselves to any policy that's designed for clowns.

THE LIFE-SAVING QUALITIES OF PIZZA

A series of Italian research studies suggest that eating pizza just might do good things for a person's health.

These benefits show up, statistically speaking and seasoned with caveats, among people who eat pizza as pizza. The delightful statistico-medico-pizza effects do not happen so much, the researchers emphasize, for individuals who eat the pizza ingredients individually.

Back in 2001, Dario Giugliano, Francesco Nappo and Ludovico Coppola, at Second University Naples, published a study in the journal *Circulation* called 'Pizza and Vegetables Don't Stick to the Endothelium'. The thrust of their finding was that, unlike many other typical Italian meals, pizza does not necessarily cause clogged blood vessels (atherosclerosis) and death.

Silvano Gallus of the Istituto di Ricerche Farmacologiche, in Milan, has cooked up several studies about the health effects of ingesting pizza.

In 2003, together with colleagues from Naples, Rome and elsewhere, Gallus published a report called 'Does Pizza Protect Against Cancer?', in the *International Journal of Cancer*. It compares several thousand people who were treated for cancer of the oral cavity, pharynx, oesophagus, larynx, colon or rectum with patients who were treated for other, non-cancer ailments. Several hospitals gathered data about what the patients said they habitually ate. The study ends up speaking, in a vague, general way of an 'apparently favorable effect of pizza on cancer risk in Italy'.

In 2004, in a monograph in the *European Journal of Cancer Prevention*, Gallus and two colleagues wrote that 'regular consumption of pizza, one of the most typical Italian foods, showed a reduced risk of digestive tract cancers.' Later that year, another team anchored by Gallus published 'Pizza and Risk of Acute Myocardial Infarction'. As you would expect from the title, its purpose was 'to evaluate the potential role of pizza consumption on the risk of acute myocardial infarction'. Gallus and his team 'suggest that pizza consumption is a favourable indicator' for preventing, or at least not causing, heart attacks.

Gallus is in no way claiming that pizza prevents all ills. A Gallus-led study called 'Pizza Consumption and the Risk of Breast, Ovarian and Prostate Cancer' appeared two years later. These types of cancer are thought to arise differently from the kinds believed to be warded off by pizza. The study puts its message bluntly: 'Our results do not show a relevant role of pizza on the risk of sex hormone-related cancers.'

The Gallus studies all hedge their bets a bit. Each says, one way or another (and here I'm paraphrasing them): 'Pizza may in fact merely represent a general indicator of the so-called "Mediterranean" diet, which has been shown to have potential health benefits.'

All of this pertains to Italian-made pizza, metabolized in Italy by Italians. No matter how accurate the scientists' interpretations turn out to be, there's no guarantee that they hold true for foreign pizza, or for any pizza eaten anywhere by foreigners, or for unmetabolized pizza.

A monograph in the journal *Traffic Injury Prevention* explains that, whatever the good or bad of eating pizza may be, delivering the pies can put you on a collision course with unhappiness.

Dr Chris McLean and his colleague J. Bernard at Mayday University Hospital, in Croydon, UK, say they were inspired by a 1992 report, in the journal *Injury*, by Dr M.G. Dorrell of Edgware General Hospital in London. Dorrell 'described a series of six patients who sustained bony injuries in road traffic accidents during the course of their employment as pizza delivery personnel'. Subsequently, the Pizza and Pasta Association, acting in concert with the government, developed a voluntary code of practice for home delivery individuals, with the goal of reducing or even eliminating pizza/transportation-induced bony and other injuries.

McLean and Bernard, a decade after the Edgeware pizza crack-up study, analysed what happened to three pizza delivery moped drivers who were themselves delivered to Mayday University Hospital. 'None of them possessed a full UK driver's license', write McLean and Bernard, and 'all three were involved in collisions with automobiles.' One simply fell off his moped; the other two 'somersaulted over their moped handlebars'. Piecing together the available evidence, McLean and Bernard tentatively conclude that 'nonnative workers who lack English language skills and moped driving skills are at increased risk of moped accidents'.

Thus, pizza can most definitely not be ruled out as a nexus of havoc prior to ingestion.

Giugliano, Dario, Francesco Nappo, and Ludovico Coppola (2001). 'Pizza and Vegetables Don't Stick to the Endothelium'. *Circulation* 104 (7): E34–5.

Gallus, Silvano, Cristina Bosetti, Eva Negri, Renato Talamini, Maurizio Montella, Ettore Conti, Silvia Franceschi, and Carlo La Vecchia (2003). 'Does Pizza Protect Against Cancer?'. *International Journal of Cancer* 107 (2): 283–4.

Gallus, Silvano, Cristina Bosetti, and Carlo La Vecchia (2004). 'Mediterranean Diet and Cancer Risk'. *European Journal of Cancer Prevention* 13 (5): 447–52.

Gallus, Silvano, A. Tavani, and Carlo La Vecchia (2004). 'Pizza and Risk of Acute Myocardial Infarction'. *European Journal of Clinical Nutrition* 58 (11): 1543–6.

Gallus, Silvano, Renato Talamini, Cristina Bosetti, Eva Negri, Maurizio Montella, Silvia Franceschi, A. Giacosa, and Carlo La Vecchia (2006). 'Pizza Consumption and the Risk of Breast, Ovarian and Prostate Cancer'. *European Journal of Cancer Prevention* 15 (1): 74–6.

McLean, C. R., and J. Bernard (2003). 'Ethnicity as a Factor in Pizza Delivery Crashes'. *Traffic Injury Prevention* 4 (3): 276–7.

Dorrell, M. G. (1992). 'The Cost of Home Delivery.' *Injury* 23 (7): 495–6.

MAY WE RECOMMEND

CANNIBALISM: ECOLOGY AND EVOLUTION AMONG DIVERSE TAXA
by M.A. Elgar and B.J. Crespi (Oxford University Press, 1992)

On page 361 the monograph gets right to the nub of the matter: 'Cannibalism is a particularly antisocial form of behaviour.'

THE SMELL OF CADAVERINE IN THE MORNING

Putrescine and cadaverine, the two most frighteningly named of all chemicals, lurk in our mouths all day, every day. This simple fact emerged in 2003 when Professor Michael Cooke BSc PhD CChem FRSC Eur Chem, of the Centre for Chemical Sciences, Royal Holloway, University of London, and two colleagues published a delectable horror story of a study called

'Time Profile of Putrescine, Cadaverine, Indole and Skatole in Human Saliva'. It appeared in the *Archives of Oral Biology*. That journal – surprisingly, given its content – is a regular haunt of only a tiny fraction of the world's horror fiction enthusiasts.

Professor Cooke and his companions imply that other chemists had grown discouraged at the prospect of doing a time profile of putrescine, cadaverine, indole and skatole in human saliva. The odour of saliva is intensely bland, compared to that of its most apallingly stenched components, and in a certain chemical sense, stable. An American group, they say, 'reported their inability to increase the odour of saliva'. But Cooke and his team gave it a go and succeeded.

In isolation, putrescine and cadaverine are anything but bland. They smell even worse than their names suggest. The one was so named because it evokes, and is involved in, the putrefaction of flesh. The other's name suggests, deliberately and accurately, the stench of rotting corpses. Biochemically, the pair are cousins, and though not inseparable, are often found in each other's company. Together with their mundane-sounding (yet also pungent) companions indole and skatole, putrescine and cadaverine are formed by 'bacterial putrefaction of saliva in the oral cavity', explains the study.

Here's how Cooke and his companions went about monitoring their presence.

Twelve dentally healthy volunteers, three women and nine men, supplied the spit. The scientists took pains to collect and handle it properly. They explain that 'The unstimulated saliva was expectorated into a glass vial coated with 5 mg of NaF [sodium fluoride] to inhibit further [chemical reactions].'

The results appear in a simple graph. Its lines and data points tell an engrossing tale: the 'mean concentration of cadaverine, putrescine and indole in saliva throughout the day'. (Skatole, though mentioned in the study's title, never appears in measurable amounts.)

Cadaverine, putrescine and indole levels, having built up overnight, are high when we awaken. But 'they are rapidly reduced by the combined action of eating breakfast and oral cleaning'. Then until the beginning of the work day (around 9:00 am) the levels remain steady.

The rest of the day has its ups and downs. Cadaverine, putrescine and indole levels rise until mid-morning, then slowly decline. Noontime brings dramatic change – the amounts are 'reduced by the mechanical action of chewing involved in the ingestion of lunch'. After lunch, the levels rise pretty steadily until it's time to knock off work.

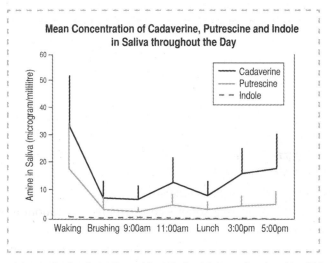

Bacterial putrefaction in twelve subjects in 'good general and oral health' who had consumed 'a controlled diet the evening before sample collection and during the saliva collection day'

The researchers stopped collecting data every day at 5:00 pm. What happens between then and dawn retains, slightly, an air of mystery.

Cooke, Michael, N. Leeves, and C. White (2003). 'Time Profile of Putrescine, Cadaverine, Indole and Skatole in Human Saliva'. *Archives of Oral Biology* 48 (4): 323–7.

THE GIANT WHO PLAYED WITH DEAD DOLLS

Frances Glessner Lee, the giant astride the world of miniature crime scenes, died nearly fifty years ago. Lee built a collection of what she called 'nutshell studies', each a tiny, high-precision recreation of a room in which a murder had been committed.

Each featured a little victim, in or on whom the wee murder weapon was embedded or enwrapped. The many lavishly grim elements of each diorama were, mostly, copped and composited from stories of real crimes.

Lee and her nutshell studies have a context. She endowed an entire, entirely new programme at Harvard Medical School: the department of legal medicine. The concocted crime scenes served as its mesmerizing centre of activity.

The authorities knew that Lee manufactured her evidence from whole cloth, sliced wallpaper, glass, wood, paint and other materials. They knew that she bankrolled the entire operation. They knew that she enlisted the aid of a carpenter, a pricey interior decorating firm and a company that makes doll's houses. They knew that she conspired with a large number of police officers, whom she plied with lavish meals and strong drink. No one entirely figured out her motive.

Lee, the heiress of a wealthy Chicago farming-equipment manufacturing family, chose the department's first (and only) leader – a dashing male doctor, her brother's Harvard classmate, whom she had kept very much in mind during the decades that preceded her inheritance.

Several times a year, she would invite and fund police officers and medical examiners from across the US, thirty or forty at a time, to travel to her Harvard seminar in homicide investigation. Everyone would examine and discuss the miniature rooms, then go and dine together in splendour at one of

Boston's finest hotels. Lee even bought the hotel a costly set of china for use exclusively at these dinners.

Author-photographer Corinne May Botz crafted a book entitled *The Nutshell Studies of Unexplained Death*. Originally published in 2004, it shows appropriately disturbing close-up photos of the artificial crime scenes. Botz also reproduces the short descriptive texts that each visiting law-enforcement official was expected to read in connection with his (they were, apparently, all men) visit to Lee and her educational programme.

The US National Library of Medicine has put several of Botz's photos online (http://www.nlm.nih.gov/visibleproofs/ galleries/biographies/lee.html). A documentary film, *Of Dolls and Murder*, came out in 2012, with creepy John Waters narrating.

Lee-style fantastically detailed miniature crime-scene recreations never became a standard tool for crime-scene investigators. But their spirit lives on. There is now something of a police vogue for crime incident diagramming software, polyflex forensic mannequins and mini tubular dowel crime-scene reconstruction kits.

The Harvard department of legal medicine did not long survive the passing of its founder and funder. Its crown jewels, the little rooms, went south and now reside at the Maryland Medical Examiner's Office in Baltimore.

Botz, Corinne May (2004). *The Nutshell Studies of Unexplained Death: Essays and Photography*. New York: Monacelli Press.

Eckert, W.G., (1981). 'Miniature Crime Scenes: A Novel Use in Crime Seminars'. *American Journal of Forensic Medicine & Pathology* 2 (4): 365–8.

Lee, Frances Glessner (1952). 'Legal Medicine at Harvard University'. *Journal of Criminal Law, Criminology, and Police Science* 42 (5): 674–8.

IN BRIEF

'SPATIAL PATTERNING OF VULTURE SCAVENGED HUMAN REMAINS'

by M. Katherine Spradley, Michelle D. Hamilton and Alberto Giordano (published in *Forensic Science International*, 2011)

The authors, at Texas State University-San Marcos, report: 'After the initial appearance of the vultures, the body was reduced from a fully-fleshed individual to a skeleton within only 5 h.'

FAST ROBUST AUTOMATED BRAIN EXTRACTION

Fast Robust Automated Brain Extraction

Stephen M. Smith[*]

Oxford Centre for Functional Magnetic Resonance Imaging of the Brain, Department of Clinical Neurology, Oxford University, John Radcliffe Hospital, Headington, Oxford, United Kingdom

Scientists marvel at how other scientists – the ones who study something other than what they themselves study – give strange meanings to common words.

Evan Shellshear, at Fraunhofer Chalmers Centre in Gothenburg, sent me an example, a study called 'Fast Robust Automated Brain Extraction'.

Shellshear said: 'I stumbled across this article somehow [whilst] looking for optimal code to quickly compute the distance between two triangles in three-dimension space for computer games. It sounds almost like something out of a game itself … After careful reading, [the paper] justifies the initially shocking title.' The author is Stephen M. Smith who, back in 2002 when the paper came out, was at the department of clinical neurology at Oxford University's John Radcliffe Hospital, and is now a professor of biomedical engineering.

Certain details might give you the willies, if you are unpractised in the ways and words of Dr Smith's line of research. Especially in this age when zombies are so much in the public mind. One section of the paper carries the conceivably disturbing headline 'Overview of the brain extraction method'.

The abstract could plausibly have been written by Dr Phibes or any of a hundred other horror-movie body-part-snatching researchers. It says: 'Brain Extraction Tool (BET) … is very fast … [I give] the results of extensive quantitative testing against "gold-standard" hand segmentations, and two other popular automated methods.'

That phrase 'hand segmentations' suggests lots of lengthy, laborious tedium. But in some contexts 'hand segmentations' could be a handy euphemism – in a discussion, say, of how to pluck out only the choicest parts of a cadaver's brain after you've smashed open its skull.

Smith acknowledges that doing his deed by hand has one advantage over letting a computer do it: 'Manual brain/non-brain segmentation methods are, as a result of the complex information understanding involved, probably more accurate than fully automated methods are ever likely to achieve.'

But he explains that, financially, it's better when a computer does the dirty work. 'There are serious enough problems with manual segmentation', he warns. 'The first problem is time cost. Manual brain/non-brain segmentation typically takes between 15 minutes and two hours.'

'Fast Robust Automated Brain Extraction' is not about sucking brains out of people's skulls, alas.

Published in a journal called *Human Brain Mapping*, it's about perceiving more clearly what's in the pictures produced by modern imaging machines. These magnificent devices give such a profusion of detail that doctors sometimes can't tell where one body part ends and another begins.

Smith explains that in a brain scan, 'the high resolution magnetic resonance image will probably contain a considerable amount [of] eyeballs, skin, fat, muscle, etc'. The image can become more understandable, more useful 'if these non-brain parts of the image can be automatically removed'.

Thus a report that gives the heebie-jeebies to some scientists gives, instead, hope and cheer to those who have the specialized brains to appreciate it.

Smith, Stephen M. (2002). 'Fast Robust Automated Brain Extraction'. *Human Brain Mapping* 17: 143–55.

IN BRIEF

'GOOD SAMARITAN SURGEON WRONGLY ACCUSED OF CONTRIBUTING TO PRESIDENT LINCOLN'S DEATH: AN EXPERIMENTAL STUDY OF THE PRESIDENT'S FATAL WOUND'
by J.K. Lattimer and A. Laidlaw (published in the *Journal of the American College of Surgeons,* 1996)

The authors, who are at the College of Physicians and Surgeons of Columbia University in New York, explain that 'When President Abraham Lincoln was shot in the back of the head at Ford's Theater in Washinton, DC, on April 14, 1865, he was immediately rendered unconscious and apneic. Doctor Charles A. Leale, an Army surgeon, who had special training in the care of brain injuries, rushed to Lincoln's assistance... thrusting his finger into the brain through the finger hole.'

DEAD MULE SPECIALITIES

Jerry Leath Mills reigns as the unchallenged authority on the subject of dead mules in twentieth-century American southern literature.

Professor Mills established his reputation – almost instantly – in 1996, with the publication of a long essay called 'Equine Gothic: The Dead Mule as Generic Signifier in Southern Literature of the Twentieth Century'. He retired that year after three decades of teaching English at the University of North Carolina at Chapel Hill. His dead mule treatise appeared in the *Southern Literary Journal*.

'Equine Gothic' reads as if the accumulated dead mules had been stewing in Mills's head, and were at last in a fit state for him to ladle out. 'My survey of around 30 prominent 20th-century southern authors', Mills writes, 'has led me to conclude … that there is indeed a single, simple, litmus-like test for the quality of southernness in literature, one easily formulated into a question to be asked of any literary text and whose answer may be taken as definitive, delimiting and final. The test is: Is there a dead mule in it?'

'Drowning – Faulkner's most commonly employed means of dispatch for the mules in his work' from 'The Dead Mule Rides Again' (drawing by Bruce Strauch)

He organized his findings 'coroner-wise', listing the different causes of southern literary mule death. These include:

'Beating'. In Larry Brown's novel Dirty Work

'Coal dust and mine gasses'. In Hubert J. Davis's short story 'The Multilingual Mule'

'Collision with railroad train'. In William Faulkner's 'Mule in the Yard'

'Drowning'. In many works by many authors

'Decapitation by irate opera singer'. In Cormac McCarthy's *The Crossing*

'Falls from cliffs'. In Cormac McCarthy's *Blood Meridian*

In Mills's telling, McCarthy is an impresario of fictional mule slaughter. In *Blood Meridian* alone, 'Mules are shot, roasted, drowned, knifed, and slain by thirst; but the largest number, fifty …, plummet from a single cliff during an ambush, performing an almost choreographic display of motion and color.'

Other mules, in other stories by other southerners, exit by freezing, hanging, gunshot wounds, a 'fall into subterranean cavity', rabies, stab wounds, 'something called vesicular stomatitis', thirst, overwork or, when all else failed to fell them, 'unspecified natural causes'.

Mills expresses wonder at all this belletristic mule mortality. During his own life lived in the American south, he says, 'I have never laid eyes on an actual dead mule.'

Through his very bookishness, he came to realize that others shared this particular obliviousness. 'I have been gratified of late to discover that I am not alone', he beams. 'I am pleased to read, in an article in *Scientific American* magazine, that the British army harbors a proverbial belief that "one never sees a dead mule".'

I must report, though, that the *Scientific American* article goes on to say about the British army: 'During World War I many men made the acquaintance of mules for the first time, and many mules had their first encounter with partially trained drivers … [This] ended only too often in events belying the tradition.'

Mills, Jerry Leath (1996). 'Equine Gothic: The Dead Mule as Generic Signifier in Southern Literature of the Twentieth Century'. *Southern Literary Journal* 29 (1): 2–17.
— (2000). 'The Dead Mule Rides Again'. *Southern Cultures* 6 (4): 11–34.
Savory, Theodore H. (1970). 'The Mule'. *Scientific American* (December): 102–9.

SOFT IS HARD

MAY WE RECOMMEND

'DOES TELEVISION ROT YOUR BRAIN? NEW EVIDENCE FROM THE COLEMAN STUDY'
by Matthew Gentzkow and Jesse M. Shapiro (National Bureau of Economic Research paper, 2006)

Some of what's in this chapter: His proud 'simplistic view of corruption' • Narcissism writ big • Dietrich's doll fetish • The further, two-dimensional, adventures of Jesus • A market for stock metaphors • Baggagial emotion • Oboe danger • The Thatcher cut-off • No yes • A scarcity of bad singers • Nervousness detector • Filthy words from the distaff side • Personality of veg • Benefits of randomness • Double-entry Australadventures • An exacting approach to vagueness • Do not mention the wolves • Wedding rings are wearing • Your discoveries

THE ECONOMIST WHO THEORIZED ON CORRUPTION

Steven Ng-Sheong Cheung, because of his adventure with the legal system – not despite it – sets a high standard for economists. The economics profession is often accused of concocting clever theories that don't resonate in the lives of real people. Cheung devised a theory about man's struggle with corruption and governments. He wrote about his theory, with relish. The US government shone a spotlight on Professor Cheung's thoughts when it issued a warrant for his arrest.

Back in 1996, Cheung – who was then head of the University of Hong Kong's School of Economics and Finance, and an economist at the University of Washington in Seattle – published 'A Simplistic General Equilibrium Theory of Corruption' in the journal *Contemporary Economic Policy*. The very first sentence of that paper says: 'The author's simplistic view of corruption is that all politicians and government officials – like everyone else – are constrained self-maximizers. They therefore establish or maintain regulations and controls with the intent to facilitate corruption, which then becomes a source of income for them.'

Cheung dives deep into the matter. A few pages later he explains: 'I made the now famous statement that it is no use to put a beautiful woman in my bedroom, naked, and ask me not to be aroused. I said that the only effective way of getting rid of corruption is to get rid of the controls and regulations that give rise to corruption opportunities.'

A press release issued on 25 February 2003 by the Seattle office of the US Department of Justice bears the headline: 'ARREST WARRANTS ISSUED FOR ECONOMIST AND WIFE FOR THEIR FAILURE TO APPEAR' (in all uppercase letters). That press release reported: 'A federal grand jury returned an indictment against the CHEUNGs on January 28, 2003. STEVEN N.S. CHEUNG was named in all thirteen counts of the indictment, charging him with Conspiracy to Defraud the United States, six counts of filing false income tax returns, and six counts of filing false foreign bank account reports.'

Previously, an investigative team at the *Seattle Times* newspaper had alleged that Cheung was linked to an art gallery that sold fake antiques.

In 2004, Steven N.S. Cheung Inc., the corporation in front of the man behind the corporation, sued the US government for 'recovery of wrongful levy' of $1,434,573.76 in taxes. Cheung's company won, but then lost a court appeal over how much money was involved.

Professor Cheung retired to China, where he became a newspaper columnist and blogger (blog.sina.com.cn/zhangwuchang), joining a profession that enjoys almost as much public confidence and respect as his former one.

Cheung, Steven Ng-Sheong (1996). 'A Simplistic General Equilibrium Theory of Corruption'. *Contemporary Economic Policy* 14 (3): 1–5.

US Department of Justice (2003). 'Arrest Warrants Issued for Economist and Wife for Their Failure to Appear'. Press release, Seattle, 25 February, http://www.justice.gov/tax/usaopress/2003/txdv03cheung.html.

US Department of Justice (2003). 'Eminent Economist and Wife Indicted on Tax Charges'. Press release, 28 January, Seattle, http://www.justice.gov/tax/usaopress/2003/txdv03cheung012803.html.

Wilson, Duff (2003). 'Economist Tied to Fake Art Faces Tax Charges'. *Seattle Times*, 29 January, http://community.seattletimes.nwsource.com/archive/?date=20030129&slug=cheung29.

US Court of Appeals for the Ninth Circuit (2008). *Steven N.S. Cheung Inc v. United States of America*, no. 07-35161, Seattle, 23 September.

IN BRIEF

'NARCISSISM IS A BAD SIGN: CEO SIGNATURE SIZE, INVESTMENT, AND PERFORMANCE'
by Charles Ham, Nicholas Seybert and Sean Wang (UNC Kenan-Flagler research paper, 2013)

The authors explain: 'Using the size of the CEO signature on annual SEC filings to measure CEO narcissism, we find that narcissism is positively associated with several measures of firm overinvestment, yet lower patent count and patent citation frequency. Abnormally high investment by narcissists predicts lower future revenues and lower sales growth. Narcissistic CEOs also deliver worse current performance as measured by return on assets, particularly for firms in early life-cycle stages and with uncertain operating environments, where a CEO's decisions are most likely to impact the firm's future value. Despite these negative performance indicators, more narcissistic CEOs enjoy higher compensation, both unconditionally and relative to the next highest paid executive at their firm.'

Samples: CEO signatures taken from annual reports

THE FETISHIZED DIETRICH'S DOLL FETISH

'The study of Marlene Dietrich's relationship with her dolls has taken me into some new research territory.' With these words, Judith Mayne, the Distinguished Humanities Professor of French Studies and Italian Studies and Women's Studies at Ohio State University, takes many of us into some new territory. Her recent study 'Marlene, Dolls, and Fetishism', published in the journal *Signs*, is an introductory guidebook to this intriguing land.

Scholars of the German film star Marlene Dietrich have debated many of the contradictions in her screen roles and in her life. Professor Mayne singles out a contradiction the other scholars have apparently failed to appreciate: 'The particular contradiction that I find the most challenging in relationship to Dietrich is that this icon of sophistication and glamour was the proud owner of a black doll, which she called her "mascot" and carried with her everywhere during her career.'

The chief source of information about the doll is a memoir that Dietrich's daughter, Maria, published in 1992. Drawing from that memoir, Mayne describes the central role that the doll had in Dietrich's life:

'Marlene came home in a fury one day, desperately digging through trunks and accusing little Maria of having stolen her doll. Maria knew that her father, Rudolf Sieber, had been fixing the grass skirt of the doll; the father is thus identified as both maternal and as the devoted handmaiden to his wife's desires. Daughter Maria concluded that the doll was Dietrich's 'good-luck charm throughout her life – her professional one'.

Mayne reports that this was not the only doll that Dietrich owned, and that this doll was not a strictly private object. 'Indeed, attentive viewers might well recognize the doll from Dietrich's films as well as from publicity postcards of the actress, often in the company of yet another one of Dietrich's dolls, a so-called Chinese coolie doll, made by Lenci.' Mayne concentrates her analytical powers on the black doll, dismissing the Chinese coolie doll as likely just 'a companion' to the other, and paying the other dolls scarcely a mention.

The black doll appeared in at least four of Dietrich's films, including the one that made her famous: *The Blue Angel*. Mayne concentrates on that film. In so doing, she gives us her highly original – and, I must say, challenging – take on Dietrich, dolls and fetishism.

'It is perhaps easy (too easy) to see the doll in *The Blue Angel* as a classic fetish', Mayne writes. 'Fetishism, understood now more as the ambivalence of the ironist than the dread of the phallocrat, is, if not celebrated, then at the very least explored as offering an understanding of multiple identifications and positions … [But] the racial dynamics of Dietrich's relationship with her dolls foreclose any simple celebration of fetishism as necessarily subversive.'

Mayne, Judith (2004). 'Marlene, Dolls, and Fetishism'. *Signs* 30 (1): 1257–63.

IN BRIEF

'THE ORGANIZATION OF SANTA: FETISHISM, AMBIVALENCE AND NARCISSISM'
by Robert Cluley (published in *Organization*, 2011)

JESUS, FLESHED OUT

The Tring tiles add spice, and maybe a little sugar, to the question, 'What would Jesus do?' These adventure-packed ceramic cartoons depict scenes, unmentioned in the bible, from Jesus's boyhood, and appear to be based on the Apocryphal Infancy of Christ Gospels, 'which purport to tell events from Jesus' life, from ages 5 – 12'. Apparently created in the fourteenth century, they once enlivened a wall of the parish church in the town of Tring, though, according to some accounts, 'the Infancy stories were so startling that Church Fathers condemned them as unsuitable for inclusion in the canon'.

Several of the tiles now reside under glass, in Case 15 of Room 40 of the British Museum in London. Every year, almost all of the museum's five million or so visitors pass them by, unaware of the potent lessons that the tiles offer.

The spare New Testament description of Jesus gets some fleshing out. The museum has provided labels that help the visitor make sense of the action. (The museum changes the labels now and again; the descriptions below are the wording as it appeared when I first encountered the tiles.)

We see young Jesus learning how to play with others, and how not to. The accompanying label explained: 'Jesus plays by the side of the river Jordan making pools; a boy destroys one with a stick and falls dead.' Then, 'The Virgin admonishes Jesus, who restores the boy to life by touching him with his foot.'

We see that Jesus the boy is rambunctiously playful: 'A schoolmaster is seated on the left. Before him stands Jesus, a boy leaping on to his back in attack. The boy falls dead.' Then 'Two women complain to Joseph on the left, while Jesus restores the boy to life.'

As with all artwork, religious or otherwise, these images invite further, and perhaps different, interpretations. As the boy is falling dead, the schoolmaster and Jesus rather appear to be smiling and giving each other what is now known as a 'high five'. The two women who complain about the death do so with what would, in other circumstances, be taken as smug smiles.

Where the museum labels' author observed that Jesus revives the other boy – the one who mucked with Jesus's pools by the river Jordan – 'by touching him with his foot', some may think they see a kick being administered.

A few of the cartoon sequences are incomplete. For these, the museum's captions shared what happens in the missing tiles.

We see the curiosity-driven Jesus learn, in fits and starts, some fine points about how to teach people a lesson: 'Parents, to prevent their children playing with Jesus, have shut them in an oven; Jesus, asking what that oven contains, is told "Pigs". (The remainder of the story, showing the children transformed into pigs and their subsequent cure by Jesus, is missing.)'

We see how, as word about the boy Jesus spreads, neighbours sometimes leap to unkind assumptions about the lad: 'A father locks his boy up in a tower to stop him playing with Jesus.' Then 'Jesus helps the boy out of the tower. (The remainder of this story, showing the father struck blind on his return, is missing.)'

The Tring tiles presumably will continue their residence in the British Museum. There are no announced plans to reproduce them for contemplative use in churches, offices or the home.

Bagley, Ayers (1985). 'Jesus at School'. *Journal of Psychohistory* 13 (1): 13–31.
Casey, Mary F. (2007). 'The Fourteenth-century *Tring Tiles*: A Fresh Look at Their Origins and the Hebraic Aspects of the Child Jesus' Actions'. *Peregrinations* 2 (2): n.p., http://peregrinations.kenyon.edu/vol2-2/current.html.

MAY WE RECOMMEND

'JESUS AND THE IDEAL OF THE MANLY MAN IN NEW ZEALAND AFTER WORLD WAR ONE'
by Geoffrey M. Troughton (published in *Journal of Religious History*, 2006)

THE DEAD CAT BOUNCE

In struggling to make sense of the stock market, people reach and stretch for metaphors. Sometimes they even contort, dislocate and mangle. In 1995, Geoff P. Smith of the University of Hong Kong made a grand unified effort to gather and classify those metaphors.

Smith congealed the metaphors and his thoughts into a monograph called 'How High Can a Dead Cat Bounce?: Metaphor and the Hong Kong Stock Market'. It appeared in the journal *Hong Kong Papers in Linguistics and Language Teaching*.

Resident Fauna

The Hong Kong stock market, as other stock markets of the world, is inhabited by two well-known metaphorical animals, the bull and the bear. These creatures have been in existence for a long time, and the *Oxford English Dictionary* notes that the word "bear" was in use in the early eighteenth century, and common around the time of the great speculative frenzy known as the South Sea Bubble. The origin of the term "bear" is probably connected to the proverb "to sell the bearskin before one has caught the bear". A bear is now understood as a person who predicts a fall in the price of stocks, but originally it appears to have referred to the stocks themselves, as in the phrase "to sell the bear". This would nowadays be referred to as

Detail: 'Resident Fauna' of 'How High Can a Dead Cat Bounce?'

Smith collected mostly from three sources: the *South China Morning Post*'s business supplement, the *Asian Wall Street Journal* and the *Asia Business News* television programme.

Here are verbatim snippets, which I present in the form of a mixed-metaphor story. It begins, of course, with two favourite animals.

> Strong bears came out of the woods determined to drag the market down.
> The bears had their claws firmly dug in and were not letting go.
> Optimists saw the makings of a baby bull, but naysayers warned it could be a bum steer ... after last year's grizzly bear market.
> Speculators played a cat and mouse game with stocks.
> The stock remained a dog.
> Investors [ran] like a herd of startled gazelles.
> The market was very nervous.

The market was having trouble focusing on issues.

Sick dollar … groggy dollar … dollar cringes.

The market was suffering vertigo.

The market started to drift and lose direction.

[The market] precariously balanced on the 10,050 mark.

The index hovered.

[The market was] losing its footing.

The index fell off the cliff.

The Hang Seng Index dropped like a brick. [This one's
    a simile. I know, I know.]

[The] index continued its tailspin.

The market seemed to have come out of its free fall.

Stock prices took a roller coast ride and ended up in
    the subway.

The bounce was more technical than substantial.

Those hoping for a big rebound to catapult it out of
    this bear trap would probably be disappointed.

The question every trader will be asking himself this
    week is: just how high can a dead cat bounce?

Surveying the hotchpotch of stock phrases and market-driving hype, Smith sighs: 'Rarely do commentators say "These events are totally unpredictable; I haven't the slightest idea what caused them to occur." '

The possibility flaunts itself that no one quite understands what the stock market's doing. If that's the case, everyone's unlikely to come up with metaphors that truly fit. But that won't stop them from trying. Smith tells us why at the end of his report: 'A group with a significant stake in the maintenance of an impression of certainty are the financial "gurus" whose words and actions can have profound effects on the way markets move … To a lesser extent, a host of commentators, analysts and advisers benefit from the illusion that market events are controlled and rational and can be explained and predicted.'

Smith, Geoff P. (1995). 'How High Can a Dead Cat Bounce?': Metaphor and the Hong Kong Stock Market'. *Hong Kong Papers in Linguistics and Language Teaching* 18: 43–57.

IN BRIEF

'ON WHAT I DO NOT UNDERSTAND (AND HAVE SOMETHING TO SAY), PART I'
by Saharon Shelah (published in *Fundamenta Mathematicae*, 2000)

AN ACADEMIC VIEW ON EMOTIONAL BAGGAGE

Until 1997, lost luggage just sat there, ignored, while scholars focused on other subjects. Then Klaus R. Scherer and Grazia Ceschi of the University of Geneva went to an airport and took a hard look at the emotions engendered by luggage loss.

Scherer and Ceschi used hidden cameras, microphones and survey forms to record people's reactions to learning that their luggage was lost. Their report, called 'Lost Luggage: A Field Study of Emotion–Antecedent Appraisal', published in the journal *Motivation and Emotion*, concerns 112 luggage-deprived passengers. It looks at several questions.

> **QUESTION ONE**: When airline passengers discover that their baggage is lost, which emotions do they feel, and with what intensity?

The researchers focused on five basic emotions: anger, resignation, indifference worry and good humour. They tried to measure the intensity of each of these twice – first before and then after the passengers spoke with an airport baggage agent. The study presents the data in a series of tables and graphs, which culminate in an image labelled 'Component loadings in a three-dimensional solution produced by nonlinear canonical correlation of appraisal dimensions and emotion ratings'.

QUESTION TWO: Can the emotional reactions reported by these airline passengers help prove or disprove one of the fundamental tenets of a theory called 'Appraisal Theory'?

Specifically, Scherer and Ceschi want to know what these luggage-loss emotions can tell us about 'the notion that objectively similar events or situations will elicit dissimilar patterns of emotional reactions in different individuals, due to variable appraisal patterns'. It's a subtle question, but Scherer and Ceschi find a simple, clear answer. Appraisal Theory, they conclude, is not yet up to snuff. It will need some tinkering before it can adequately explain how airline passengers react after learning that their luggage has gone missing.

QUESTION THREE: Once a baggage agent enters the picture, how does he or she affect the passenger's luggage-loss-related emotions?

This is the question they ask, and it may be an important question, but Scherer and Ceschi are really after something deeper. What they really want to know, they say, is what happens in the end. What emotion will passengers feel after they have (a) lost their luggage, and then (b) talked with an airport baggage agent, and then (c) decided how well (or how badly) the whole baggage retrieval service worked?

The answer, say Scherer and Ceschi, is that people 'tended to be less angry and/or sad if they were satisfied with the performance of the retrieval service', or if the airport baggage agent made 'a positive impression' on them.

The researchers were surprised by the strong extent of what they call 'emotion blending'. The unfortunate passengers often felt more than pure anger or resignation or worry. Sometimes they felt both anger *and* resignation, or both anger *and* worry.

Scherer and Ceschi welcome such unexpected discoveries. And so they offer a suggestion for their fellow social scientists. Actually seeing how people behave, they write, can 'widen the perspective'. It's a 'corrective against the self-insulation tendencies' of relying too much on theories about how people behave.

Scherer, Klaus R., and Grazia Ceschi (1997). 'Lost Luggage: A Field Study of Emotion-Antecedent Appraisal'. *Motivation and Emotion* 21 (3): 211–35.
— (2000). 'Criteria for Emotion Recognition from Verbal and Nonverbal Expression: Studying Baggage Loss in the Airport'. *Personality and Social Psychology Bulletin* 26 (3): 327–39.

OBOE, SOLO

'Dangerous Trends in Oboe Playing' burst into print in 1973, a fiery warning for all to see.

Melvin Berman, then a professor of oboe and chamber music at the University of Toronto, daringly discussed – openly, in public – a problem that threatened his tradition-bound minority community. Publishing in the *Journal of the International Double Reed Society*, he wrote: 'The American oboist is being branded as an in-bred, dull and insensitive technician by most of the world's orchestral musicians and many of the world's finest conductors ... Something is desperately wrong and everyone, except the oboists, knows what it is. Now, I think it's time for the oboists to find out.'

This was the roiling societal ferment of the 1960s, at last and after a delay of several years reaching into even the tiniest and most isolated of social groups. It was also a *cri de coeur*: 'We have forgotten, or refuse to accept the fact that there are other schools of playing, other approaches to the oboe, other methods of making reeds, which deserve and have at least as wide an acceptance as our own ... We are taught to laugh at the English vibrato, smirk at the German sound, ridicule the French brightness.' The solution, insisted Professor Berman, was 'to liberate ourselves from the restrictive attitudes imposed upon us by certain elements of the past generation'.

The Berman article appeared at a peculiar moment. The oboe community had just been confronted and transfixed by a challenge to its deepest values. This, of course, was the appearance of the Italian composer Luciano Berio's *Sequenza VII*, a piece written for solo oboe. So novel was it that two decades passed before a scholar, Cason A. Duke of Louisiana State University and Agricultural and Mechanical College, described it coherently: 'a single note held as a drone by the audience throughout the piece'. (Berio later reworked the composition as *Sequenza VIIb* for soprano saxophone.)

The oboists were not unique in feeling isolated. That same issue of the *Journal of the International Double Reed Society* also contained a revealing study by the Austrian bassoonist Hugo Burghauser. He described the loneliness of bassoonists, and also the physically painful reality of their existence.

Burghauser's writing is intensely personal: 'I was often asked why I chose to play the bassoon. The answer is simple and might be the same for many other bassoonists: in my youth, students for this instrument were very scarce so when I wanted to enter the Vienna Conservatory the Dean of this prominent Academie offered me every course free of all tuition if only I would take up the bassoon.'

Burghauser – and his fellows' – pain shines through in passages such as this one: 'Many years ago also, there was the experience in Europe of manufacturing reeds from steel. They sounded beautiful, spoke easily and lasted of course for a very long time, with but one hitch – after five minutes playing a splitting headache resulted!'

Berman, Melvin (1973). 'Dangerous Trends in Oboe Playing'. *Journal of the International Double Reed Society* 1 (1): n.p.

Burghauser, Hugo (1973). 'Philharmonic Adventures with Bassoon'. *Journal of the International Double Reed Society* 1 (1): n.p.

Berman, Melvin (1988). *Art of Oboe Reed Making*. Toronto: Canadian Scholars' Press.

Duke, Cason A. (2001). 'A Performer's Guide to Theatrical Elements in Selected Trombone Literature'. Thesis, School of Music, Louisiana State University and Agricultural and Mechanical College, May.

THATCHER, INTERRUPTED

Prime Minister Margaret Thatcher's masterful way of handling interruptions inspired one psychologist to study, intently, how she did it. As this scholar communicated his findings to the public, other scholars, with different views, interrupted him – and he them.

Geoffrey Beattie is a professorial research fellow of the Sustainable Consumption Institute at the University of Manchester. Or rather, was – the 12 November 2012 issue of the *Manchester Evening News* quotes 'a university spokesman' as saying: 'We can confirm Geoff Beattie has been dismissed from the university for gross misconduct following a disciplinary hearing.' In 1982, however, while he was at the University of Sheffield, Beattie published two studies about Thatcher.

The first, in the journal *Semiotica*, says: 'Mrs Thatcher's interviews display a distinctive pattern – she is typically interrupted more frequently than other senior politicians, and she is interrupted more often by interviewers than she herself interrupts … Butting-in interruptions … occur where her interviewer interrupts but fails to get the floor.'

> Butting-in interruptions are particularly common in her interviews. These occur where her interviewer interrupts but fails to get the floor, as in the following example where Denis Tuohy (D.T.) tries to take the floor from Mrs Thatcher (M.T.) (*TV Eye*, April 1979).
>
> M.T.:   ... if you've got the money in your pocket you can
>   choose/whether you spend it on things which attract
>   Value Added Tax/or not
> D.T.:   (You s-
>   (
> M.T.:   (and the main necessities don't
> D.T.:   You say a little on Value Added Tax
>
> where   /indicates unfilled pause ≥200 ms
>
>   (word 1
>   (word 2 indicates simultaneous speech
>
> A series of experiments was designed to test the hypothesis that Mrs Thatcher is often interrupted because she displays turn-yielding cues at points where she has not completed her

From 'Why Is Mrs Thatcher Interrupted So Often?'

The second study, called 'Why Is Mrs Thatcher Interrupted So Often?', appears in the journal *Nature*. Beattie (and two colleagues) analysed how Thatcher deployed her voice, words and gaze when Denis Tuohy interviewed her on the television programme *TV Eye*. The study explains that most people use fairly standard cues with each other, as to when each will stop or start talking. But, it explains: 'Many interruptions in an interview with Mrs Margaret Thatcher, the British prime minister, occur at points where independent judges agree that her turn appears to have finished. It is suggested that she is unconsciously displaying turn-yielding cues at certain inappropriate points.'

Peter Bull and Kate Mayer of the University of York disagreed. They and Beattie argued, slowly, in public, mostly in the pages of the *Journal of Language and Social Psychology*. In 1988, Bull and Mayer wrote a treatise called 'Interruptions in Political Interviews: A Study of Margaret Thatcher and Neil Kinnock', saying their 'results were quite contrary to what might have been expected from the work of Beattie'. Beattie replied with 'Interruptions in Political Interviews: A Reply to Bull and Mayer'.

Then, a year later, Bull and Mayer countered with 'Interruptions in Political Interviews: A Reply to Beattie'. And Beattie responded with 'Interruptions in Political Interviews: The Debate Ends?'.

Bull and Mayer, four years later, published a study called 'How Not to Answer Questions in Political Interviews'. After that, the conversation dwindled.

But Beattie did not shy away from studying interruptive behaviour. He wrote a book called *On The Ropes: Boxing as a Way of Life*. A reviewer, in the journal *Aggressive Behavior*, contended that the book is a knuckle-cracking good read, that it 'takes you into a world where respect has nothing to do with your publication record or letters after your name, but with your ability to take a punch without letting on that you are hurt. Geoffrey gives a wonderful description of what it is like to be on the receiving end of an accomplished sparring partner.'

Beattie, Geoffrey W. (1982). 'Turn-taking and Interruption in Political Interviews: Margaret Thatcher and Jim Callaghan Compared and Contrasted'. *Semiotica* 29 (1/2): 93–114.

—, Anne Cutler and Mark Pearson (1982). 'Why Is Mrs Thatcher Interrupted So Often?'. *Nature* 300 (23/30 December): 744–7.

Bull, Peter, and Kate Mayer (1988). 'Interruptions in Political Interviews: A Study of Margaret Thatcher and Neil Kinnock'. *Journal of Language and Social Psychology* 7 (1): 35–46.

Beattie, Geoffrey (1989). 'Interruptions in Political Interviews: A Reply to Bull and Mayer'. *Journal of Language and Social Psychology* 8 (5): 327–39.

Bull, Peter, and Kate Mayer (1989). 'Interruptions in Political Interviews: A Reply to Beattie'. *Journal of Language and Social Psychology* 8 (5): 341–4.

Beattie, Geoffrey (1989). 'Interruptions in Political Interviews: The Debate Ends?'. *Journal of Language and Social Psychology* 8 (5): 345–8.

Bull, Peter, and Kate Mayer (1993). 'How Not to Answer Questions in Political Interviews'. *Political Psychology* 14 (4): 651–66.

Beattie, Geoffrey (1996). *On the Ropes: Boxing as a Way of Life*. London: Victor Gollancz.

Archer, John (1997). 'Book Review: *On the Ropes: Boxing as a Way of Life,* by Geoffrey Beattie'. *Aggressive Behavior* 23 (3): 215–6.

RESEARCH SPOTLIGHT

'THE DEMISE OF "YES": AN INFORMAL LOOK'

by John W. Trinkaus (published in *Perceptual and Motor Skills,* 1997)

Perceptual and Motor Skills, 1997, 84, 866.   © Perceptual and Motor Skills 1997

THE DEMISE OF "YES": AN INFORMAL LOOK[1]

JOHN TRINKAUS

College of Business Administration, St. John's University

Summary.—For affirmative responses to simple interrogatories, the use of "absolutely" and "exactly" may be becoming more socially frequent than "yes." A counting of positive replies to 419 questions on several TV networks showed 249 answers of "absolutely," 117 "exactly," and 53 of "yes."

The author explains: 'For affirmative responses to simple interrogatories, the use of "absolutely" and "exactly" may be becoming more socially frequent than "yes". A counting of positive replies to 419 questions on several TV networks showed 249 answers of "absolutely", 117 "exactly", and 53 of "yes".'

HARMONIOUS RESULTS

Despite what you may have heard, acoustical analysis suggests that (1) most people are not horrible singers, and (2) most horrible singers are not tone deaf – they're just horrible singers.

In 2007, Isabelle Peretz and Jean-François Giguère of the University of Montreal, and Simone Dalla Bella, of the University of Finance and Management in Warsaw, tested the abilities of sixty-two 'nonmusicians' in Quebec who admitted to being 'occasional singers'.

The very act of testing proved more difficult than Dalla Bella, Giguère and Peretz had heard it would be. Still, they published their report, in the *Journal of the Acoustical Society of America*, called 'Singing Proficiency in the General Population'. They complain, to anyone who will listen, that it is not easy to measure the goodness or badness of singing. There is 'no consensus', they seem to wail, 'on how to obtain … objective measures of singing proficiency in sung melodies'.

They devised their own test, using the refrain from a song called 'Gens du Pays', which people in Quebec commonly sing as part of their ritual to celebrate a birthday. That refrain, they explain, has thirty-two notes, a vocal range of less than one octave, and a stable tonal centre.

These scientists went to a public park, where they used a clever subterfuge to recruit test subjects: 'The experimenter pretended that it was his birthday and that he had made a bet with friends that he could get 100 individuals each to sing the refrain of Gens du Pays for him on this special occasion.'

The resulting recordings became the raw material for intensive computer-based analysis, centring on the vowel sounds – the 'i' in 'mi', for example. Peretz, Giguère and Dalla Bella assessed each performance for pitch stability, number of pitch interval errors, 'changes in pitch directions relative to musical notation', interval deviation, number of time errors, tempo-ral variability and timing consistency. For comparison, they recorded and assessed several professional singers performing the same snatch of song.

Peretz, Giguère and Dalla Bella give a cheery assessment of the untrained, off-the-street singers: 'We found that the major-ity of individuals can carry a tune with remarkable proficiency. Occasional singers typically sing in time, but are less accurate in pitch as compared to professional singers. When asked to slow down, occasional singers greatly improve in performance, making as few pitch errors as professional singers.' Only a very few, they say, were 'clearly out of tune'.

The scientists then focused on two of the horrid singers. Both screechers were 'aware that they sang out of tune'. They proved to be almost the opposite of tone deaf. When tested, they 'correctly detected 90% and 96% of pitch deviations in a melodic context'.

Peretz has continued to study the mystery of poor sing-ing. In 2012, she and her colleague Sean Hutchins finished a comprehensive study. Harmoniously with the earlier finding, they conclude that most bad singers are good at hearing fine sounds, and bad at making them.

Dalla Bella, Simone, Jean-François Giguère and Isabelle Peretz (2007). 'Singing Proficiency in the General Population'. *Journal of the Acoustic Society of America* 121 (2): 1182–9.
Hutchins, Sean, and Isabelle Peretz (2012). 'A Frog in Your Throat or in Your Ear?: Searching for the Causes of Poor Singing'. *Journal of Experimental Psychology: General* 141 (1): 76–97.

NERVOUS CALLS

Businesses need a way to detect nervousness on the telephone, says a 2011 patent, which offers a computerised means of accom-plishing this.

Inventor Valery Petrushin obtained his doctorate in computer science from the Glushkov Institute for Cybernetics, Kiev, and now works in Illinois. His patent is for a method of 'detecting emotion in voice signals in a call center'.

A simple flow chart illustrates 'a method for detecting nervousness in a voice in a business environment to prevent fraud'. We see the following three statements, each enclosed in its own box: 'Receiving voice signals from a person during a business event'; 'Analyzing the voice signals for determining a level of nervousness of the person during the business event'; 'Outputting the level of nervousness of the person prior to completion of the business event'. Petrushin recommends his invention also to improve 'contract negotiation, insurance dealings … in the law enforcement arena as well as in a courtroom environment, etc'. He specifies a few of the many applications in these fields: 'fear and anxiety could be detected in the voice of a person as he or she is answering questions asked by a customs officer, for example'.

Petrushin assigned his patent rights to Accenture Global Services Ltd, the giant international advice-for-almost-everything consulting company. Accenture sells call-centre-centric services to BSkyB ('BSkyB increased its customer satisfaction while enhancing its bottom-line performance', its website once noted) and other high-flying firms.

The patent, in essence, presents a recipe that has missing steps. That's because scientists have not yet found a reliable mechanical way to identify emotions. But there's hope (the document implies), in that 'psychologists have done many experiments and suggested theories'.

The method relies on 'statistics of human associations of voice parameters with emotions'. These parameters are acoustical – all about the vibrations of the voice, paying no attention to the words those sounds happen to represent. Words can be misleading; if spoken different ways, they might carry different emotions.

This all grew out of, and in a sense headed sideways from, Petrushin's early research about telephone-borne emotion. Those were the days before fear came to prominence. A study he published in 2000 identified a different emotion as the one to focus on for industrial purposes. Petrushin wrote at that time: 'It was not a surprise that anger was identified as the most important emotion for call centers.'

People Performance Confusion Matrix

Category	Normal	Happy	Angry	Sad	Afraid
Normal	**66.3**	2.5	7.0	18.2	6.0
Happy	11.9	**61.4**	10.1	4.1	12.5
Angry	10.6	5.2	**72.2**	5.6	6.3
Sad	11.8	1.0	4.7	**68.3**	14.3
Afraid	11.8	9.4	5.1	24.2	**49.5**

After recording one of four sentences: 'This is not what I expected'; 'I'll be right there'; 'Tomorrow is my birthday'; and 'I'm getting married next week'. One of the study question posed: 'Which kinds of emotions are easier/harder to recognize?' The researchers explain: 'people better understand how to express/decode anger and sadness than other emotions'.

His experiment back then measured the ability of then-current computer programs to correctly identify different emotions in recorded voices. The patent filing, eleven years later, speaks of improvement in the emotional technology. In the best of several test runs, it says, 'We can see that the accuracy for fear is higher (25–60%) … The accuracy for sadness and anger is very high: 75–100% for anger and 88–93% for sadness.'

Petrushin, Valery A. (2011). 'Detecting Emotion in Voice Signals in a Call Center'. US patent no. 7,940,914, 10 May.

— (2000). 'Emotion Recognition in Speech Signal: Experimental Study, Development, and Application'. *Proceedings of the Sixth International Conference on Spoken Language Processing*, Beijing: 222–5.

MAY WE RECOMMEND

'THE PERSONALITY OF VEGETABLES: BOTANICAL META-
PHORS FOR HUMAN CHARACTERISTICS'
by Robert Sommer (published in *Journal of Personality*, 1988)

WOMEN'S DIRTY WORDS

'Expletives of Lower Working-Class Women', published in 1992
in the journal *Language in Society*, is a rare sociolinguistic study
of this inherently provocative topic. 'This article', wrote author
Susan Hughes of the University of Salford, 'sets out to look at
the reality of the swearing used by a group of women from a
deprived inner-city area'. Hughes surveyed six women in Ordsall, a part of Salford said to be characterized by 'social malaise'.

'My observations of these women', Hughes wrote, 'showed
me that, contrary to some theories, they use a strong vernacular
style ... These women are proud of their swearing: "We've taught
men to swear, foreigners what's come in the pub." Their general
conversation is peppered with fuck, twat, bastard, and so on.
Yet they do differentiate between using swearwords in general
conversation and using them with venom and/or as an insult.'

That traditional practice of womanly swearing, if it is to
survive, must deal with challenges. Fifteen years after the Hughes
study was published, a corrupting influence came to town. A
home insurance firm announced that Ordsall had attained a place
on the company's Young Affluent Professionals Index. Ordsall,
they said, had become a 'property hotspot' attracting wealthy
young professionals. The newcomers will have something to
say about the community's evolving expletive standards.

Hughes' study gave scholars a clean picture of those
standards prior to yuppie adulteration. She perhaps hinted
that tensions would arise if outsiders were to move in. 'The use
of "prestigious" standard English has no merit nor relevance

for these [lower working-class] women, it cannot provide any social advantage to them or increase any life chances for them. In fact, the standard norm would isolate them from their own tight-knit community.'

Hughes explicity based her inquiry on an elegant, simple piece of research performed a few years earlier. Barbara Risch, at the University of Cincinnati, published a study called 'Women's Derogatory Terms for Men: That's Right, "Dirty" Words'. Risch surveyed forty-four female, mostly middle-class students, asking each of them to answer the following question:

> There are many terms that men use to refer to women which women consider derogatory or sexist (broad, chick … piece of ass, etc.). Can you think of any similar terms or phrases that you or your friends use to refer to males?

Risch reported that: 'A classification system for the fifty variant terms emerged quite naturally from the data obtained. The responses can be classified under the following headings: references to birth, ass, head, dick, boys, animal, meat, and other.'

She gave advice to future expletive researchers: 'The importance of female interviewers for the results of this study cannot be overemphasized. It is doubtful whether any response would have been elicited in the presence of male interviewers.'

Recent expletive research uses MRI scanners to try to see what happens physically in someone's brain when they swear. Typified by 1999's 'Expletives: Neurolinguistic and Neurobehavioral Perspectives on Swearing', these studies owe a debt, at least in spirit, to the cussing women of Ordsall and Cincinnati.

Hughes, Susan E. (1992). 'Expletives of Lower Working-Class Women'. *Language in Society* 21 (2): 291–303.

Risch, Barbara (1987). 'Women's Derogatory Terms for Men: That's Right, "Dirty" Words'. *Language in Society* 16 (3): 353–8.

Van Lancker, Diana, and Jeffrey L. Cummings (1999). 'Expletives: Neurolinguistic and Neurobehavioral Perspectives on Swearing'. *Brain Research Reviews* 31 (1): 83–104.

IN BRIEF

'USAGE AND ORIGIN OF EXPLETIVES IN BRITISH ENGLISH'
by Hana Cechová (thesis, Masaryk University Brno, Czech Republic, 2006)

PARTY POLITICS GONE RANDOM

Democracies would be better off if they chose some of their politicians at random. That's the word, mathematically obtained, from a team of Italian physicists, economists and political analysts. The team includes a trio whose earlier research showed, also mathematically, that bureaucracies would be more efficient if they promoted people at random.

Alessandro Pluchino, Andrea Rapisarda, Cesare Garofalo and two other colleagues at the University of Catania in Sicily published their new study in a physics journal *Physica A: Statistical Mechanics and its Applications*. The study itself is titled 'Accidental Politicians: How Randomly Selected Legislators Can Improve Parliament Efficiency'.

The scientists made a simple calculation model that mimics the way modern parliaments work, including the effects of particular political parties or coalitions. In the model, individual legislators can cast particular votes that advance either their own interests (one of which is to gain re-election), or the interests of society as a whole. Party discipline comes into play, affecting the votes of officials who got elected with help from their party.

But when some legislators are selected at random – owing no allegiance to any party – the legislature's overall efficiency improves. That higher efficiency, the scientists explain, comes in 'both the number of laws passed and the average social welfare obtained' from those new laws.

Parliamentary voting behaviour echoes, in a surprisingly

detailed mathematical sense, something economist Carlo M. Cipolla sketched in his 1976 essay published in book form, *The Basic Laws of Human Stupidity* (see page 9). Cipolla gave an insulting, yet possibly accurate, description of any human group: 'human beings fall into four basic categories: the helpless, the intelligent, the bandit and the stupid'. Pluchino, Rapisarda, Garofalo and their colleagues base their mathematical model partly on this fourfold distinction.

The maths indicate that parliaments work best when some – but not all – of the members have been chosen at random. The study explains how a country, subject to the quirks of its own system, can figure out what mix will give the best results.

Random selection may feel like a mathematician's wild-eyed dream. It's not. The practice was common in ancient Greece, when democracy was young. The study tells how, in Athens, citizens' names were placed into a randomisation device called a kleroterion.

Later on, legislators were selected randomly in other places, too. In Bologna, Parma, Vicenza, San Marino, Barcelona and bits of Switzerland, say the scientists, and 'in Florence in the 13th and 14th century and in Venice from 1268 until the fall of the Venetian Republic in 1797, providing opportunities to minorities and resistance to corruption'.

Athens, way back when, used random selection to people its juries. So, still, does much of the world.

And it's not just juries. Iceland, having survived a financial collapse in the first decade of the twenty-first century, set about devising a new constitution. For advice on that, the nation assembled a committee of 950 citizens chosen at random.

In 2010, Pluchino et al. were awarded an Ig Nobel Prize in management for demonstrating mathematically that organizations would become more efficient if they promoted people at random.

Pluchino, Alessandro, Cesare Garofalo, Andrea Rapisarda, Salvatore Spagano and Maurizio Caserta (2011). 'Accidental Politicians: How Randomly Selected Legislators Can Improve Parliament Efficiency'. *Physica A: Statistical Mechanics and its Applications* 390 (21/22): 3944–54.

Pluchino, Alessandro, Caesar Garofalo and Andrea Rapisarda (2010). 'The Peter Principle Revisited: A Computational Study'. *Physica A: Statistical Mechanics and its Applications* 389 (3): 467–72.

Cipolla, Carlo M. (1976). *The Basic Laws of Human Stupidity*. Bologna: The Mad Millers/Il Mulino.

GREAT ADVENTURES IN ACCOUNTING

The supposedly staid, unglamorous field of accounting is in fact packed, to some degree, with exciting adventures. But accountants rarely divulge this fact to persons outside the profession. Three monographs, all produced in Australia, document some of the adventure – and even some of the excitement.

In 1967, a paper by Professor R.J. Chambers of the University of Sydney, called 'Prospective Adventures in Accounting Ideas', appeared in the journal *Accounting Review*. Looking both backwards and forwards, Chambers enthuses ruefully: 'These fifty years have seen quite a few potentially fruitful ideas, with wide implications, brought to notice, noticed scarcely at all and almost abandoned … Some 43 years ago, [accounting scholar Henry Rand] Hatfield said "Let us boldly raise the question whether accounting, the late claimant for recognition as a profession, is not entitled to some respect, or must it consort with crystal-gazing … and palmreading?" I wonder what Hatfield would think today, to see how far some would have us go in the direction of crystal-gazing. I leave you to think about what I am referring to.'

Decades of accounting adventures later, Lee D. Parker of the University of Adelaide penned a thirty-one-page study, 'Historiography for the New Millennium: Adventures in Accounting and Management', for a rival journal. Parker explains, almost giddily, that 'we stand on the threshold of the new millennium, facing, on one hand, an academy avowedly presentist and futurist in its research orientation, but, on the other hand, signs of a society rediscovering its past with upsurges of interest in heritage building and artefacts preservation, cinema audiences attracted to the plots of Thomas Hardy and Jane Austen and a host of historical periods and events, historical tourism, thriving antique furniture markets and so on.'

Parker comes to a barely restrained conclusion: 'While we owe the twentieth century founders of accounting and management history a considerable debt, there remains much for us to learn and even more to discover. Let us begin.'

The turn of the century brought a new openness to, maybe even nostalgia and yearning for, accounting adventure, symbolized by the publication of a jaunty paper by Lorne Cummings and Mark Valentine St. Leon of Macquarie University in the journal *Accounting History*.

Capital Investment for 'Road' Shows

Year	Company	Investment	AU$ 2005
1855	Brown's	AU £3,000	122,000
1883	St Leon's	AU £5,000	298,000
1896	Probasco's	AU £3,000	229,000
1907	Wirth Bros	AU £15,000	2,175,508
1912	Bud Atkinson's	AU £8,300	481,000
1935	Ivan Bros'	Au £40,000	1,682,000
1948	Wirth Bros'	AU £250,000	5,889,632
1961	Bullen Bros'	AU £250,000	2,668,319

Table 1 from 'Jugglers, Clowns and Showmen: The Use of Accounting Information in Circus in Australia'. The authors state: 'Bemoaning the liquidation of his Wild West and circus enterprise in Melbourne in 1913, the American showman Bud Atkinson said that "about 8,300 pounds [was] put into the show", a figure which probably includes the cost of shipping personnel from the west coast of the United States, as well as the purchase, or construction, and shipment of its American-built wagons.'

Called 'Juggling the Books: The Use of Accounting Information in Circus in Australia', it savours the accounting practices of Australian circuses during the period from 1847 to 1963. 'Responding to the call for an increased historical narrative in accounting', write Cummings and St. Leon, 'we have studied the literature, documentation and personal memoirs concerning circus in Australia ... We have established that, despite elementary levels of education, many circus people exhibited an intuitive grasp of fundamental accounting principles, albeit in a rudimentary form. Nevertheless, since financial and management reporting practises were typically unsystematic, and even non-existent, in all but the largest circus enterprises, Australian circus management may not have been optimized.'

Chambers, R.J. (1967). 'Prospective Adventures in Accounting Ideas'. *Accounting Review* 42 (2): 241–53.
Parker, Lee D. (1999). 'Historiography for the New Millennium: Adventures in Accounting and Management'. *Accounting History* 4 (2): 11–42.
Cummings, Lorne, and Mark Valentine St. Leon (2009). 'Juggling the Books: The Use of Accounting Information in Circus in Australia'. *Accounting History* 14 (1-2): 11–33.

MAY WE RECOMMEND

'VAGUENESS: AN EXERCISE IN LOGICAL ANALYSIS'
by Max Black (published in *Philosophy of Science*, 1937)

SHEPHERDS AND HUNTERS AND BUTCHERS, OH MY

When you hobnob with Slovakian shepherds, don't mention wolves. A new study called 'Mitigating Carnivore-Livestock Conflict in Europe: Lessons from Slovakia', says: 'Compared to other sectors of society shepherds had the most negative attitudes, particularly towards wolves.'

Wolves are again roaming the forests of Slovakia. They were almost wiped out in the mid-twentieth century, then reappeared thanks to a thirty-year moratorium on hunting. Now the small-but-growing wolf population has restored its tradition of helping local livestock go missing or get mauled.

The researchers, Robin Rigg and Maria Wechselberger at the Slovak Wildlife Society, Slavomír Findo at the Carpathian Wildlife Society, Martyn Gorman at the University of Aberdeen, and Claudio Sillero-Zubiri and David Macdonald at the University of Oxford, published their work in the journal *Oryx*.

They found that, mostly, wolves grab sheep near the edge of a forest, especially if the shepherds employ what the scientists call 'ineffective methods (chained dogs and inadequate electric fencing)'. Experimenting, the team identified two effective methods: unchained guard dogs and adequate fencing. As with much research, the obvious was apparently not obvious beforehand to all who needed to know about it.

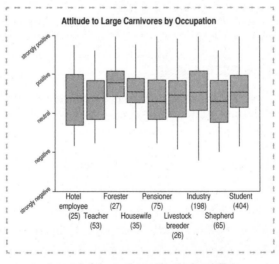

Appreciating carnivore-livestock conflict

That's the story with shepherds. Now, for people who kill lots of animals: hunters and butchers. A study by an Austrian/

British/Malaysian team probes the psychological differences between them.

The monograph, 'Multi-method Personality Assessment of Butchers and Hunters: Beliefs and Reality', appeared last year in the journal *Personality and Individual Differences*. The authors, Martin Voracek (the one and same, mentioned elsewhere in this volume), Stefan Stieger and Viren Swami, learned, through direct questioning, that 102 Austrian university students feel hunters and butchers have 'higher aggressiveness and masculinity'. The students also seem to believe that hunters – but not butchers – possess unusually high self-esteem.

The researchers then studied twenty-five hunters and twenty-three butchers from rural Lower Austria, and compared them with forty-eight persons who neither hunt nor butcher.

First, they tested for hypermasculinity, using a survey technique that 'gauges macho personality'. They verified their findings 'unobtrusively', by doing an analysis based on the relative lengths of each person's second and fourth fingers.

Then they used a standard test to measure each person's self-esteem. They double-checked by having each individual rate 'the likeability of all letters of the alphabet', bearing in mind that 'letters appearing in individuals' names, especially their initials, are rated more favorably than the remainder of the alphabet'.

All told, the researchers 'found little evidence for the factuality of' hunters or butchers having great masculinity, aggressiveness or self-esteem.

Rigg, Robin, Salvomír Findo, Maria Wechselberger, Martyn L. Gorman, Claudio Sillero-Zubiri and David W. Macdonald (2011). 'Mitigating Carnivore-Livestock Conflict in Europe: Lessons from Slovakia'. *Oryx* 45 (2): 272–80.

Voracek, Martin, Daniela Gabler, Carmen Kreutzer, Stefan Stieger, Viren Swami and Anton K. Formann (2010). 'Multi-method Personality Assessment of Butchers and Hunters: Beliefs and Reality'. *Personality and Individual Differences* 49 (7): 819–22.

RESEARCH SPOTLIGHT

'DIGIT RATIO (2D:4D) AND WEARING OF WEDDING RINGS'
by Martin Voracek (published in *Perceptual and Motor Skills*,
2008)

WEDDING RINGS, DIMINISHED OR DISMISSED

A gold wedding band symbolises permanence, but bits of it disappear as a marriage endures, scraping against the marital skin every moment that metal and finger convene. Georg Steinhauser, a chemist at Vienna University of Technology, calculated how much goes missing, how quickly and at what cost.

Figure 1: 'The author's wedding ring' from Steinhauser (2008)

Steinhauser's study,' Quantification of the Abrasive Wear of a Gold Wedding Ring', appears in a 2008 issue of *Gold Bulletin*, a quarterly journal published by the World Gold Council, whose stated goal is 'to stimulate desire for gold by articulating core truths and discovering new opportunities'.

Steinhauser got married. A week later he weighed his wedding ring. He weighed it every week over the next year. The report shows a graph of the weight, revealing an average loss of about 0.12 milligrams per week.

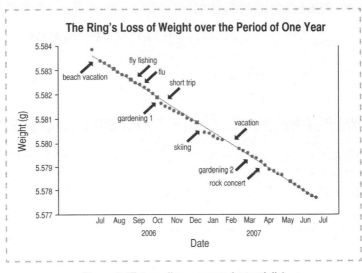

Figure 2: 'Extraordinary events (potentially)
influencing the abrasion are noted'.

Steinhauser estimated that, every year, the city of Vienna, with slightly more than 300,000 married couples, suffers an aggregate loss from its rings of about 2.2 kilograms of eighteen-carat gold, worth (at the time) approximately 35,000 euros.

Steinhauser says: 'Due to abrasion of metal particles, human fingers wearing gold rings leave a trace of gold almost everywhere.' He warns his fellow scientists to 'not wear gold rings in analytical laboratories that are dedicated to the analysis of traces of metals, because a gold ring or the skin that has been in contact with the ring are possible sources of contamination'.

So it is important to give some thought to where you wear your wedding ring, a fact brought bleakly home, and then to an infirmary, in 2002. The urology department of the University of Bonn in Germany received a visit from a fifty-nine-year-old man who had slipped his wedding band from its habitual home on a knuckled digit on to a different, non-knuckled digit, to which it became tightly attached. Too tightly.

After a period of questioning and photograph-taking, the medical staff brought out a small machine called a metal-ring cutter. They snipped the ring asunder, freeing it from the withered post that had come to inhabit it in a fashion known medically as 'strangled'.

A quartet of medicos published a graphic study about this in the journal *Urology*. 'To our knowledge', they write, 'this is the first report of a wedding ring used as [a] constriction device.'

Steinhauser, Georg (2008). 'Quantification of the Abrasive Wear of a Gold Wedding Ring'. *Gold Bulletin* 41 (1): 51–7.
Perabo, Frank G.E., Gabriel Steiner, Peter Albers and Stefan C. Müller (2002). 'Treatment of Penile Strangulation Caused by Constricting Devices'. *Urology* 59 (1): 137.

CALL FOR SUBMISSIONS

If you know of any improbable research – the sort that makes you laugh and think, and that you think will make other people laugh, too – I would be delighted and grateful to hear about it.

Please email me at marca@improbable.com with an improbable subject line of your choosing.

ACKNOWLEDGEMENTS

Thanks to my wife, Robin, our socially challenged yet ever-helpful dog Milo, my parents, my ever-astounding editor, Robin Dennis, and equally stout-hearted agents, Regula Noetzli and Caspian Dennis (who is still, as far as I know, not related by blood or marriage to Robin Dennis).

Thanks to helpful friends/colleagues at *The Guardian*, especially Tim Radford, Will Woodward, Claire Phipps, Donald MacLeod, Alice Wooley, Ian Sample and Alok Jha.

Thanks to each of the many people who told me about things that wound up in this book, some of whom are named: Richard Akerman, Gábor Andrássy, Claudio Angelo, Catherine L. Bartlett, Yoram Bauman, Sandra Betar, Tommy Burch, Jim Cowdery, Sofia Dahl, Kristine Danowski, Tommy Dighton, Martin Eiger, Stefanie Friedhoff, Martin Gardiner, Eric Geigle, Tom Gill, Jessica Girard, Diego Golombeck, Richard Gustafson, Stephen Hale, Andrea Halpern, N. Hammond, Silvia Haneklaus, Kathryn Hedges, Mark Henderson, D.E. Hepplewhite, K. Kris Hirst, Torbjörn Karfunkel, Geoffrey Kendrick, Maarten Keulemans, Greg Kohs, Erwin Kompanje, C. Lajcher, Frederic Lepage, Mark Lewney, El Lisse, Shelly Marino, Pedro Marques-Vidal, Benno Meyer-Rochow, James H. Morrissey, Ernst Niebur, Mason Porter, Hanne Poulsen, Tim Reese, Achim Reisdorf, G. Jules Reynolds, Felicia Sanchez, Ewald Schnug, Ivan Schwab, Sally Shelton, Derek R. Smith, John Troyer, Kurt Verkest, Marcy Weisberg, Greg Wells, Anna Wexler, and quite a few more whose names I apologize for not listing here.

Thanks also to the following individuals who have helped in delightful, and sometimes improbable ways, in recent years:

Siobhan Abeyesinghe, Dany Adams, Henry Akona, Faraz Alam, Kendra Albert, Deborah Anderson, Helen Arney, Amar Asher, James Bacon, Richard Baguley, Robin Ball, Chris Balliro, Matthew Battles, Bob Batty, Jackie Baum, Alice Bell, Jim Bell, Michael Berry, Monica Berry, Stephan Bolliger, Tina Bowen, Jim Bredt, Ryan Budish, Estrella Burgos, Heidi Clark, Brian Clegg, Charlotte Burn, Mary Carmichael, Rita Carter, Sarah Castor-Perry, Sylvie Coyaud, J.V. Chamary, Stuart Clark, Julie Clayton, Brian Clegg, Stevyn Colgan, Michael Conterio, Mo Costandi, Ian Day, Charles Deeming, Neil Denny, Judith Donath, Nick Doody, Gary Dryfoos, Chris Dunford, Stanley Eigen, the family Eliseev/Eliseeva, Steve Farrar, Lucy Feilen, Maria Ferrante, Melissa Franklin, Hayley Frend, Andrew J.T. George, Jean Berko Gleason, Ray Goldstein, Alain Goriely, David C. Green, Katherine Griffin, Boris Groysberg, Jenny Gutbezahl, James Harkin, Margaret Harris, Hunter Heinlen, Jeff Hermes, Dudley Herschbach, Arthur Hinds, Jens Holbach, Adam Holland, Andrew Holding, Fariba Houman, Jonnie Hughes, Gareth Jones, Terry Jones, Susan Kany, David Kessler, Sandra Klemm, Bart Knols, Eliza Kosoy, Peter Lamont, Frederic Lepage, Maggie Lettvin, Henry Leitner, Julia Lunetta, Georgia Lyman, Mark Lynas, L. Mahadevan, Alice Martelli, Lauren Maurer, Chris McManus, Xiao-Li Meng, Thomas Michel, Kees Moeliker, Lauren Mulholland, Gustav Nilsonne, Ivan Oransky, Amanda Palmer, Rohit Parwani, Charles Paxton, Bruce Petschek, Johan Pettersson, Tacye Phillipson, Mason Porter, Thomas Povey, Aarathi Prasad, Tim Radford, Gus Rancatore, Ros Reid, Rich Roberts, SciCurious, Sid Rodrigues, Santi Rodriquez, the family Rosenberg, Steffen Ross, Michael Rutter, Faina Ryvkin, Molly Sauter, Helen Scales, Dan Schreiber, Margo Seltzer, Wolter Seuntjens, Alom Shaha, Annette Smith, Chris Smith, Alicia Solow-Niederman, Volker Sommer, Graham Southorn, Hari Sriskantha, Naomi Stephen, Richard Stephens, Brian Sullivan, Geri Sullivan, Wolter Suntjens, Vaughn Tan, Peaco Todd, Patrick Warren,

Simon Watt, Magnus Whalberg, Corky White, Tony White-head, Anna Wilkinson, Stuart Wilson, Jane Winans, Richard Wiseman, George Wolford, Thomas Woolley, Helen Zaltzman, Jonathan Zittrain, Eric Zuckerman, and quite a few more whose names I also apologize for not listing here.

EXTRA CITATIONS

RESEARCH SPOTLIGHT

Bean, Robert Bennett (1919). 'The Weight of the Leg in Living Men'. *American Journal of Physical Anthropology* 2 (3): 275–82.

Brase, Gary L., Dan V. Caprar, and Martin Voracek (2004). 'Sex Differences in Response to Relationship Threats in England and Romania'. *Journal of Social and Personal Relationships* 21 (6): 763–78.

Di Noto, Paula M., Leorra Newman, Shelley Wall and Gillian Einstein (2012). 'The Hermunculus: What Is Known about the Representation of the Female Body in the Brain?'. *Cerebral Cortex* 23 (5): 1005–13.

Einstein, Andrew J. (2010). 'President Obama's Coronary Calcium Scan'. *Archives of Internal Medicine* 170 (13): 1175–6.

Einstein, Gillian, April S. Au, Jason Klemensberg, Elizabeth M. Shin and Nicole Pun (2012). 'The Gendered Ovary: Whole Body Effects of Oophorectomy'. *Canadian Journal of Nursing Research* 44 (3): 7–17.

Einstein, Shlomo Stan. (2012). 'An Ode to Substance Use(r) Intervention Failure(s): SUIF'. *Substance Use and Misuse* 47 (13/14): 1687–721.

Manning, John T., and Peter E. Bundred (2000). 'The Ratio of 2nd to 4th Digit Length: A New Predictor of Disease Predisposition?'. *Medical Hypotheses* 54 (8): 5–7.

— and Frances M. Mather (2004). 'Second to Fourth Digit Ratio, Sexual Selection, and Skin Colour'. *Evolution and Human Behavior* 25 (1): 38–50.

Romans, Sarah, Rose Clarkson, Gillian Einstein, Michele Petrovic and Donna Stewart (2012). 'Mood and the Menstrual Cycle: A Review of Prospective Data Studies'. *Gender Medicine* 9 (5): 361–84.

Trinkaus, John W. (1980). 'Preconditioning an Audience for Mental Magic: An Informal Look'. *Perceptual and Motor Skills* 51 (1): 262.

— (1980). 'Honesty at a Motor Vehicle Bureau: An Informal Look'. *Perceptual and Motor Skills* 51 (3): 1252.

— (1990). 'Queasiness: An Informal Look'. *Perceptual and Motor Skills* 70 (2): 1393–4.

— (1997). 'The Demise of "Yes": An Informal Look'. *Perceptual and Motor Skills* 84 (3): 866.

Van den Bergh, Bram, and Siegfried Dewitte (2006). 'Digit Ratio (2D:4D) Moderates the Impact of Sexual Cues on Men's Decisions in Ultimatum Games'. *Proceedings of the Royal Society B: Biological Sciences* 273 (1597): 2091–5.

Voracek, Martin, and Maryanne L. Fisher (2002). 'Sunshine and Suicide Incidence'. *Epidemiology* 13 (4): 492–4.

— and Gernot Sonneck (2002). 'Solar Eclipse and Suicide'. *Americna Journal of Psychiatry* 159 (7): 1247–8.

Voracek, Martin A., Albinas Bagdonas, and Stefan G. Dressler (2007). 'Digit Ratio (2D:4D) in Lithuania Once and Now: Testing for Sex Differences, Relations with

Eye and Hair Color, and a Possible Secular Change'. *Collegium Antropologicum* 31 (3): 863–8.

Voracek, Martin (2008). 'Digit Ratio (2D:4D) and Wearing of Wedding Rings' *Perceptual and Motor Skills* 106 (3): 883–90.

— and Stefan G. Dressler (2010). 'Relationships of Toe-length Ratios to Finger-length Ratios, Foot Preference, and Wearing of Toe Rings'. *Perceptual and Motor Skills* 110 (1): 33–47.

Westerhof, Danielle (2005). 'Celebrating Fragmentation: The Presence of Aristocratic Body Parts in Monastic Houses in Twelfth- and Thirteenth-Century England'. *Citeaux Commentarii Cistercienses* 56 (1): 27–45.

IN BRIEF

Appel, Markus (2011). 'A Story about a Stupid Person Can Make You Act Stupid (or Smart): Behavioral Assimilation (and Contrast) as Narrative Impact'. *Media Psychology* 14 (2): 144–67.

Bernhard, Jeffrey D. (1995). 'The Swedish Pimple: Or, Thoughts on Specialization'. *Journal of the American Academy of Dermatology* 32 (3): 505–9.

Bessa Jr, Octavio. (1993). 'Tight Pants Syndrome: A New Title for an Old Problem and Often Encountered Medical Problem'. *Archives of Internal Medicine* 153 (11): 1396.

Beveridge, Allen (1966). 'Images of Madness in the Films of Walt Disney'. *Psychiatric Bulletin* 20 (10): 618–20.

Bohigian, George H. (1997). 'The History of the Evil Eye and Its Influence on Ophthalmology, Medicine and Social Customs'. *Documenta Ophthalmologica* 94 (1/2): 91–100.

Burton, Thomas C. (1998). 'Pointing the Way: The Distribution and Evolution of Some Characters of the Finger Muscles of Frogs'. *American Museum Novitates* paper no. 3229.

Cechová, Hana (2006). 'Usage and Origin of Expletives in British English'. Undergraduate thesis, Department of English Language and Literature, Masaryk University Brno, Czech Republic, 20 April.

Cluley, Robert (2011). 'The Organization of Santa: Fetishism, Ambivalence and Narcissism'. *Organization* 18 (6): 779–94.

Davidson, T.K. (1972). 'Panti-Girdle Syndrome'. *British Medical Journal* 2 (5810): 407.

Dreber, Anna, Christer Gerdes and Patrik Gränsmark (2013). 'Beauty Queens and Battling Knights: Risk Taking and Attractiveness in Chess'. *Journal of Economic Behavior and Organization* 90: 1–18.

Drummond, J.R., and G.S. McKay (1999). 'Biting Off More Than You Can Chew: A Forensic Case Report'. *British Dental Journal* 187 (9): 466.

George, M., and J. Round (2006). 'An Eiffel Penetrating Head Injury'. *Archives of Disease in Childhood* 91 (5): 416.

Ham, Charles, Nicholas Seybert and Sean Wang (2013). 'Narcissism is a Bad Sign: CEO Signature Size, Investment, and Performance'. University of North Carolina Kenan-Flagler Business School research paper no. 2013-1, 19 March.

Hansen, Christian Stevns, Louise Holmsgaard Faerch and Peter Lommer Kristensen (2010). 'Testing the Validity of the Danish Urban Myth that Alcohol Can Be Absorbed Through Feet: Open Labelled Self Experimental Study'. *British Medical Journal* 341: c6812.

Hauber, Mark E., Paul W. Sherman and Dóra Paprika (2000). 'Self-referent Phenotype Matching in a Brood Parasite: The Armpit Effect in Brown-headed Cowbirds (*Molothrus ater*)'. *Animal Cognition* 3 (2): 113–7.

Hostetter, Autumn B., Martha W. Alibali and Sotaro Kita (2007). 'Does Sitting on Your Hands Make You Bite Your Tongue? The Effects of Gesture Prohibition on Speech During Motor Descriptions'. In McNamara, D.S., and J.G. Trafton (eds). *Proceedings of the 29th Annual Conference of the Cognitive Science Society*, Nashville, TN, 1-4 August. Austin, TX: Cognitive Science Society: 1097–1102.

Hurd, Peter L., Patrick J. Weatherhead and Susan B. McRae (1991). 'Parental Consumption of Nestling Feces: Good Food or Sound Economics?'. *Behavioral Ecology* 2 (1): 69–76.

Isaacs, D. (1992). 'Glue Ear and Grommets'. *Medical Journal of Australia* 156 (7): 444–5.

Khairkar, Praveen, Prashant Tiple and Govind Bang (2009). 'Cow Dung Ingestion and Inhalation Dependence: A Case Report'. *International Journal of Mental Health and Addiction* 7 (3): 488–91.

Lattimer, J.K., and A. Laidlaw (1996).'Good Samaritan Surgeon Wrongly Accused of Contributing to President Lincoln's Death: An Experimental Study of the President's Fatal Wound'. *Journal of the American College of Surgeons* 182 (5): 431–48.

Meyer-Rochow, Victor Benno, and Jozsef Gal (2003). 'Pressures Produced When Penguins Pooh: Calculations on Avian Defaecation'. *Polar Biology* 27: 56–8.

Morse, Donald R. (1993). 'The Stressful Kiss: A Biopsychosocial Evaluation of the Origins, Evolution, and Societal Significance of Vampirism'. *Stress Medicine* 9 (3): 181–99.

Nikolić, Hrvoje (2008). 'Would Bohr Be Born If Bohm Were Born Before Born?'. *American Journal of Physics* 76 (2): 143.

Pekár, S., D. Mayntz, T. Ribeiro and M.E. Herberstein (2010). 'Specialist Ant-eating Spiders Selectively Feed on Different Body Parts to Balance Nutrient Intake'. *Animal Behaviour* 79 (6): 1301–6.

Ros, S.P., and F. Cetta (1992). 'Metal Detectors: An Alternative Approach to the Evaluation of Coin Ingestions in Children?'. *Pediatric Emergency Care* 8 (3): 134–6.

Sahoo, Manash Ranjan, Anil Kumar Nayak, Tapan Kumar Nayak and S. Anand (2013). 'Fracture Penis: A Case More Heard about than Seen in General Surgical Practice'. *BMJ Case Reports*.

Schmidt, Steven P., H. Gibbs Andrews and John J. White (1992). 'The Splenic Snood: An Improved Approach for the Management of the Wandering Spleen'. *Journal of Pediatric Surgery* 27 (8): 1043–4.

Shafik, A. (1996). 'Effect of Different Types of Textiles on Male Sexual Activity'. *Archives of Andrology* 37 (2): 111–5.

Shelah, Saharon (2000). 'On What I Do Not Understand (and Have Something to Say), Part I'. *Fundamenta Mathematicae* 168 (1/2): 1–82.

Shravat, B.P., and S.N. Harrop (1995). 'Broken Hand or Broken Nose: A Case Report'. *Journal of Accident and Emergency Medicine* 12: 225–6.

Slay, Robert D. (1986). 'The Exploding Toilet and Other Emergency Room Folklore'. *Journal of Emergency Medicine* 4 (5): 411–4.

Spradley, M. Katherine, Michelle D. Hamilton and Alberto Giordano (2011). 'Spatial Patterning of Vulture Scavenged Human Remains'. *Forensic Science International* 219 (1): 57–63.

Taylor, W.S., and E. Culler (1929). 'The Problem of the Locomotive-God'. *Journal of Abnormal and Social Psychology* 24 (3): 342–99.

Turner, M. (1991). 'Panty Hose-Pants Disease'. *American Journal of Obstetrics and Gynecology* 164 (5): 1366.

Walmsley, Christopher W., Peter D. Smits, Michelle R. Quayle, Matthew R. McCurry,

Heather S. Richards, Christopher C. Oldfield, Stephen Wroe, Phillip D. Clausen and Colin R. McHenry (2013).'Why the Long Face?: The Mechanics of Mandibular Symphysis Proportions in Crocodiles'. *PLos ONE* 8 (1): e53873.

Walz, F.H., M. Mackay and B. Gloor (1995). 'Airbag Deployment and Eye Perforation by a Tobacco Pipe'. *Journal of Trauma* 38 (4): 498–501.

Whitaker, J.H. (1977). 'Arm Wrestling Fractures: A Humerus Twist'. *American Journal of Sports Medicine* 5 (2): 67–77.

White, P.D. (1973). 'The Tight-Girdle Syndrome'. *New England Journal of Medicine* 288 (11): 584.

Williams, S. Taylor (2012). ' "Holy PTSD, Batman!": An Analysis of the Psychiatric Symptoms of Bruce Wayne'. *Academic Psychiatry* 36 (3): 252–5.

Zerbe, K.J. (1985). ' "Your Feet's Too Big": An Inquiry into Psychological and Symbolic Meanings of the Foot'. *Psychoanalytic Review* 72 (2): 301–14.

MAY WE RECOMMEND

Beaver, Bonnie V., Margaret Fischer and Charles E. Atkinson (1992). 'Determination of Favorite Components of Garbage by Dogs'. *Applied Behaviour Science* 34 (1): 129–36.

Bemelman, M., and E.R. Hammacher (2005). 'Rectal Impalement by Pirate Ship: A Case Report'. *Injury Extra* 36: 508–10.

Black, Max (1937). 'Vagueness: An Exercise in Logical Analysis'. *Philosophy of Science* 4 (4): 427–55.

Bond, J.H., and M.D. Levitt (1979). 'Colonic Gas Explosion: Is a Fire Extinguisher Necessary?'. *Gastroenterology* 77 (6): 1349–50.

Christen, J.A., and C.D. Litton (1995). 'A Bayesian Approach to Wiggle-Matching'. *Journal of Archaeological Science* 22 (6): 719–25.

Craig, J.M., R. Dow and M. Aitken (2005). 'Harry Potter and the Recessive Allele'. *Nature* 436 (7052): 776.

Eley, Karen A., and Daljit K. Dhariwal (2010). 'A Lucky Catch: Fishhook Injury of the Tongue'. *Journal of Emergencies, Trauma, and Shock* 3 (1): 92–3.

Elgar, M.A., and B.J. Crespi (eds) (1992). *Cannibalism: Ecology and Evolution among Diverse Taxa*. Oxford: Oxford University Press.

Farrier, Sarah, Iain A. Pretty, Christopher D. Lynch and Liam D. Addy (2011). 'Gagging during Impression Making: Techniques for Reduction'. *Dental Update* 38 (3): 171–6.

Fujihira, H. (1978). 'Necessary and Sufficient Conditions for Bang-Bang Control'. *Journal of Optimization Theory and Applications* 25 (4): 549–54.

Fujimura, Jun, Kenji Sasaki, Tsukasa Isago, Yasukimi Suzuki, Nobuo Isono, Masaki Takeuchi and Motohiro Nozaki (2007). 'The Treatment Dilemma Caused by Lumps in Surfers' Chins'. *Annals of Plastic Surgery* 59 (4): 441–4.

Gentzkow, Matthew, and Jesse M. Shapiro (2006). 'Does Television Rot Your Brain? New Evidence from the Coleman Study'. National Bureau of Economic Research working paper 12021, February.

Ghanem, Hussein, Sidney Glina, Pierre Assalian and Jacques Buvat (2013). 'Position Paper: Management of Men Complaining of a Small Penis Despite an Actually Normal Size'. *Journal of Sexual Medicine* 10 (1): 294–303.

Gifford, Robert and Robert Sommer (1968). 'The Desk or the Bed?'. *Personnel and Guidance Journal* 46 (9): 876–8.

Goldberg, R.L., P.A. Buongiorno and R.I. Henkin (1985). 'Delusions of Halitosis'. *Psychosomatics* 26 (4): 325–7, 331.

Hansen, Kim V., Lars Brix, Christian F. Pedersen, Jens P. Haase and Ole V. Larsen (2004). 'Modelling of Interaction between a Spatula and a Human Brain'. *Medical Image Analysis* 8 (1): 23–33.

Ho, Luis C., Alexei V. Filippenko and Wallace L.W. Sargent (1997). 'The Influence of Bars on Nuclear Activity'. *Astrophysical Journal* 487 (2): 591–602.

Jackson, C.R., B. Anderson and B. Jaffray (2003). 'Childhood Constipation Is Not Associated with Characteristic Fingerprint Patterns'. *Archives of Disease in Childhood* 88 (12): 1076–7.

Kompanje, Erwin J.O. (2013). 'Real Men Wear Kilts: The Anecdotal Evidence that Wearing a Scottish Kilt Has Influence on Reproductive Potential – How Much Is True?'. *Scottish Medical Journal* 58 (1): e1–5.

Lee, C.T., P. Williams, and W.A. Hadden (1999). 'Parachuting for Charity: Is It Worth the Money? A 5-Year Audit of Parachute Injuries in Tayside and the Cost to the NHS'. *Injury* 30 (4): 283–7.

Leung, A.K. (1984). 'Doll Shoes: The Cause of Behavioural Problems?'. *Canadian Medical Association Journal* 131 (10): 1193.

Lim, Erle C.H., Amy M.L. Quek and Raymond C.S. Seet (2006). 'Duty of Care to the Undiagnosed Patient: Ethical Imperative, or Just a Load of Hogwarts?'. *Canadian Medical Association Journal* 175 (12): 1557–9.

Lisi, Antony Garrett (2007). 'An Exceptionally Simple Theory of Everything'. arXiv.org, 6 November, http://arxiv.org/abs/0711.0770.

Mandell, Arnold J., Karen A. Selz, John Aven, Tom Holroyd and Richard Coppola (2010). 'Daydreaming, Thought Blocking and Strudels in the Taskless, Resting Human Brain's Magnetic Fields'. Paper presented at the International Conference on Applications in Nonlinear Dynamics, *AIP Conference Proceedings* 1339 (2011): 7–22.

Moeliker, C.W. (2007). 'A Penis-shortening Device Described by the 13th Century Poet Rumi'. *Archives of Sexual Behavior* 36 (6): 767.

Morice, Alyn H. (2004). 'Post-Nasal Drip Syndrome: A Symptom to be Sniffed At?'. *Pulmonary Pharmacology and Therapeutics* 17 (6): 343–5.

Petrof, Elaine O., Gregory B. Gloor, Stephen J. Vanner, Scott J. Weese, David Carater, Michelle C. Daigneault, Eric M. Brown, Kathleen Schroeter and Emma Allen-Vercoe (2013). 'Stool Substitute Transplant Therapy for the Eradication of *Clostridium difficile* Infection: "RePOOPulating" the Gut'. *Microbiome* 1 (3): http://www.microbiomejournal.com/content/1/1/3/additional.

Pon, Brian, and Alan Meier (1993). 'Hot Potatoes in the Gray Literature'. Recent Research in the Building Energy Analysis Group, Lawrence Berkeley Laboratory, paper no. 3, October.

Ryter, Stefan W., Danielle Morse and Augustine M.K. Choi (2004). 'Carbon Monoxide: To Boldly Go Where NO Has Gone Before'. *Science STKE: Signal Transduction Knowledge Environment* 2004 (230): re6.

Seimens, H.W. (1967). 'The History of Freckles in Art'. *Der Hautarzt* 18 (5): 230–2.

Silverberg, Jesse L., Matthew Bierbaum, James P. Sethna and Itai Cohen (2013). 'Collective Motion of Humans in Mosh and Circle Pits at Heavy Metal Concerts'. *Physical Review Letters* 110 (22): 228701.

Smith, G.C.S., and J.P. Pell (2003). 'Parachute Use to Prevent Death and Major Trauma Related to Gravitational Challenge: Systematic Review of Randomised Controlled Trials'. *British Medical Journal* 327: 1459–61.

THIS IS IMPROBABLE TOO

312

Sommer, Robert (1988). 'The Personality of Vegetables: Botanical Metaphors for Human Characteristics'. *Journal of Personality* 56 (4): 665–83.
Soussignan, Robert, Paul Sagot and Benoist Schaal (2007). 'The "Smellscape" of Mother's Breast: Effects of Odor Masking and Selective Unmasking on Neonatal Arousal, Oral and Visual Response'. *Developmental Psychobiology* 49 (2): 129–38.
Stern, Ronald J., and Henry Wolkowicz (1991). 'Exponential Nonnegativity on the Ice Cream Cone'. *SIAM: Journal on Matrix Analysis and Applications* 12 (1): 160–5.
Troughton, Geoffrey M. (2006). 'Jesus and the Ideal of the Manly Man in New Zealand after World War One'. *Journal of Religious History* 30 (1): 45–60.
Yerhuham, I., and O. Koren (2003). 'Severe Infestation of a She-Ass with the Cat Flea *Ctenocephalides felis felis* (Bouche, 1835)'. *Veterinary Parasitology* 115 (4): 365–7.

AN IMPROBABLE INNOVATION

Bublitz, Rebecca J., and Annette L. Terhorst (2003). 'Human-figure display system'. US patent no. 6,601,326, granted 5 August.
Castanis, George and Thaddeus Castanis (1982). 'Producing Replicas of Body Parts', US patent no. 4,335,067, granted 15 June.
Huffman, Hugh, and Ernest J. Peck (1916). 'A New and Useful Improvement in Scarecrows'. US patent no. 1,167,502, granted 11 January.
Lowe, Henry E. (1983). 'Odor Testing Apparatus'. US patent no. 4,411,156, granted 25 October.
MacDonald, Louise S. (2011). 'Internal nostril or nasal airway sizing gauge'. US patent no. 7,998,093, granted 16 August.
McMorrow, Philip E. (1968). 'Animal Track Footwear Soles'. US patent no. 3,402,485, granted 24 September.
Shemesh, David, and Dan Forman (2004). 'Automated Surveillance Monitor of Nonhumans in Real Time'. US patent no. 6,782,847, granted 31 August.
Williams, Hermann W. (1931). 'A Safety Device for Use in Making a Landing from an Aeroplane or Other Vehicle of the Air'. US patent no. 1,799,664, granted 7 April.

CALL FOR INVESTIGATORS

Haller, Daniel G. (2013). 'A Call for Opinions'. *Gastrointestinal Cancer Research* 6 (1): 1.
Oum, Robert E., Debra Lieberman and Alison Aylward. 'A Feel for Disgust: Tactile Cues to Pathogen Presence'. *Cognition & Emotion* 25 (4): 717–25.

THE INHUMANITY OF PROZAC

Deckel, A.W. (1996). 'Behavioral Changes in *Anolis carolinensis* Following Injection with Fluoxetine'. *Behavioural Brain Research* 78: 175–82.
Dodman, Nicholas H., R. Donnelly, Louis Shuster, F. Mertens, W. Rand and Klaus Miczek (1996). 'Use of Fluoxetine to Treat Dominance Aggression in Dogs'. *Journal of the American Veterinary Medical Association* 209 (9): 1585–7.
Fehrer, S.C., Janet L. Silsby and Mohamed E. El Halawani (1983). 'Serotonergic Stimulation of Prolactin Release in the Young Turkey (*Meleagris gallopavo*)'. *General and Comparative Endocrinology* 52 (3): 400–8.
Freire-Garabal, Manuel, María J. Núñez, Pilar Riveiro, José Balboa, Pablo López, Braulio

G. Zamarano, Elena Rodrigo and Manuel Rey-Méndez (2002). 'Effects of Fluoxetine on the Activity of Phagocytosis in Stressed Mice'. *Life Sciences* 72 (2): 173–83.

Henderson, Daniel R., Darlene M. Konkle and Gordon S. Mitchell (1999). 'Effects of Serotonin Re-uptake Inhibition on Ventilatory Control in Goats'. *Respiration Physiology* 115 (1): 1–10.

Huber, Robert, Kalim Smith, Antonia Delago, Karin Asaksson and Edward A. Kravitz (1997). 'Serotonin and Aggressive Motivation in Crustaceans: Altering the Decision to Retreat'. *Proceedings of the National Academy of Sciences USA* 94 (11): 5939–42.

Khan, Izhar A., and Peter Thomas (1992). 'Stimulatory Effects of Serotonin on Maturational Gonadotropin Release in the Atlantic Croaker, *Micropogonias undulatus'. General and Comparative Endocrinology* 88 (3): 388–96.

McKay, D.M., David W. Halton, J.M. Allen and Ian Fairweather (1989). 'The Effects of Cholinergic and Serotoninergic Drugs on Motility in Vitro of *Haplometra cylindracea* ('Trematoda: Digenea)'. *Parasitology* 99 (2): 241–52.

Poulsen, E.M., V. Honeyman, P.A. Valentine and G.C. Teskey (1996). 'Use of Fluoxetine for the Treatment of Stereotypical Pacing Behavior in a Captive Polar Bear'. *Journal of the American Veterinary Medical Association* 209 (8): 1470–4.

Pryor, Patricia A., Benjamin L. Hart, Kelly D. Cliff and Melissa J. Bain (2001). 'Effects of a Selective Serotonin Reuptake Inhibitor on Urine Spraying Behavior in Cats'. *Journal of the American Veterinary Medical Association* 72 (2): 173–83.

Smith, Greg N., Joseph Hingtgen and William DeMyer (1987). 'Serotonergic Involvement in the Backward Tumbling Response of the Parlor Tumbler Pigeon'. *Brain Research* 400 (2): 399–402.

Villalba, Constanza, Patricia A. Boyle, Edward J. Caliguri and Geert J. DeVries (1997). 'Effects of the Selective Serotonin Reuptake Inhibitor Fluoxetine on Social Behaviors in Male and Female Prairie Voles (*Microtus ochrogaster)'. Hormones and Behavior* 32 (3): 184–91.

ILLUSTRATION CREDITS

Grateful acknowledgement is made to the many researchers and inventors whose work is illustrated in these pages, sometimes with illustrations, sometimes without.

A classic in the body of 2D:4D work (p. x) adapted from 'A Preliminary Investigation of the Associations between Digit Ratio and Women's Perception of Men's Dance' by Bernhard Fink, Hanna Seydel, John T. Manning and Peter M. Kappeler

'Characteristics of the External Ear' (p. 2) from 'Some Characteristics of the External Ear of American Whites, American Indians, American Negroes, Alaskan Esquimos, and Filipinos' by Robert Bennett Bean

Illustrative diagram from 'A New and Useful Improvement in Scarecrows' (p. 6) from US patent no. 1,167,502

Flowchart: 'Path Model of the Criminal Achievement Process' (p. 14) adapted from 'Mentors and Criminal Achievement' by Carlo Morselli, Pierre Tremblay and Bill McCarthy

President Washington's diary of Sunday, 11 May 1788 (p. 16) courtesy of Widener Library, Harvard University

Not very to quite happy (p. 43) adapted from 'Are the Russians as Unhappy as They Say They Are?' by Ruut Veenhoven

X-ray of a three-year-old boy's head with souvenir model (p. 51) from 'An Eiffel Penetrating Head Injury' by M. George and J. Round

Inhibitions demonstrated: A celebrity fights off a paparazzo … (p. 59) from US patent no. 8,157,396

'Suzuki and Tsushihashi's classification' scheme for prints (p. 63) adapted from 'Possibilities of Cheiloscopy' by Jerzy Karsprzak

Evidence: 'Latent lip print on cotton fabric developed using Oil Red O (powder) after 30 days' (p. 65) adapted from 'Long-lasting Lipsticks and Latent Prints' by Ana Castelló, Mercedes Alvarez, Marcos Miquel and Fernando Verdú

Fig. 18 from 'Automated Surveillance of Non-humans in Real Time' (p. 68) from US patent no. 6,782,847

'Floss dispenser with six floss cutters', one for almost every day of the week (p. 79) from US patent no. 5,826,594

Gauging small, medium and large digits in place of small, medium and large nostrils (p. 83) from US patent no. 7,998,093

Observed: 'hair-whorl phenotype' (p. 99) from 'Excess of Counterclockwise Scalp Hair-Whorl Rotation in Homosexual Men' by Amar J.S. Klar

'Walter in "walking motion" ' (p. 102) from 'New Functions and Applications of Walter, the Sweating Fabric Manikin' by Jintu Fan and Xiaoming Qian

'Figures 1 and 8' (p. 103) from US patent no. 6,601,326

'Bite, shake and twist' pressure points of three crocodile species (p. 108) adapted from 'Why the Long Face?' by Christopher W. Walmsley, Peter D. Smits, Michelle R. Quayle, Matthew R. McCurry, Heather S. Richards, Christopher C. Oldfield, Stephen Wroe, Phillip D. Clausen and Colin R. McHenry

Comparison of two pedestrian approaches (p. 114) and A pedestrian flowchart (p. 115) adapted from 'Investigating Use of Space of Pedestrians' by Taku Fujiyama

Sloths, recorded, by Nyakatura and Fischer (2010) (p. 117) from 'Three-dimensional Kinematic Analysis of the Pectoral Girdle during Upside-down Locomotion of Two-toed Sloths (*Choloepus didactylus*, Linné 1758)' by John A. Nyakatura and Martin S. Fischer

Evidence: 'A sloth inside the latrine' (p. 118) from 'Disgusting Appetite' by Eckhard W. Heymann, Camilo Flores Amasifuén, Ney Shahuano Tello, Emérita R. Tirado Herrera and Mojca Stojan-Dolar

Reviewing the effect of five wok angles and three wok sizes on MAWF (p. 122) and The 'experimental layout', as viewed from above (p. 123) adapted from 'The Effect of Wok Size and Handle Angle on the Maximum Acceptable Weights of Wok Flipping by Male Cooks' by Swei-Pi Wu, Cheng-Pin Ho and Chin-Li Yen

Sock inside, on ice (p. 125) courtesy of Lianne Parkin, Sheila M. Williams and Patricia Priest

One proposed meaning of the finger (p. 133) adapted from 'Finger-length Ratios and Sexual Orientation' by Terrance J. Williams, Michelle E. Pepitone, Scott E. Christensen, Bradley M. Cooke, Andrew D. Huberman, Nicholas J. Breedlove, Tessa J. Breedlove, Cynthia L. Jordan and S. Marc Breedlove

Figure: 'Device for measuring bite force needed for breaking a toe' (p. 136) from 'The Mystery of the Missing Toes' by Bart Vervust, Stefan Van Dongen, Irena Grbac and Raoul Van Damme

The patented animal track footwear soles, compared to Kodiak bear cub tracks (p. 140) from US patent no. 3,402,485

'Experimental setup to cause tendon laceration in cadaver hands' (p. 142) from 'The Safety of Pumpkin Carving Tools' by Alexander M. Marcus, Jason K. Green and Frederick W. Werner (2004)

Analysing graphene derived from a Girl Scout cookie, chocolate, dog faeces and a cockroach (p. 146) adapted from 'Growth of Graphene from Food, Insects, and Waste' by Gedeng Ruan, Zhengzong Sun, Zhiwei Peng and James M. Tour

Fig. 1 of 15 from 'Regulation of Eating Habits' (p. 170), US patent no. 7,437,195

Steps for forming novel dry milk pieces (p. 174) and Drawing: 'a quantity of prepared cereal ...' (p. 175) adapted from US patent no. 5,827,564

Handwriting, when sober (top) and under the influence (bottom) (p. 176) from 'Handwriting Changes Under the Effect of Alcohol' by Faruk Acolu and Nurten Turan

Midsagittal image of anatomy taken during experiment 12 (p. 184) from 'Magnetic Resonance Imaging of Male and Female Genitals during Coitus and Female Sexual Arousal' by Willibrord Wijmar Schultz, Pek van Andel, Isa Sabelis and Eduard Mooyarrt

Form and cast finger (p. 204) from US patent no. 4,335,067

Penguano (p. 210) adapted from 'Pressures Produced When Penguins Pooh' by Victor Benno Meyer-Rochow and Jozsef Gal

Figure 1 (p. 215) adapted from Russian patent no. 2,399,858

A half cup of beans was consumed daily during the study period (p. 218) adapted from 'Perceptions of Flatulence from Bean Consumption among Adults in 3 Feeding Studies' by Donna M. Winham and Andrea M. Hutchins

The apparatus in action (p. 227) from US patent no. 4,411,156

Events leading to the use of the pop-up device (p. 249) from US patent no. 2005/0028720

Bacterial putrefaction in twelve subjects in 'good general and oral health' (p. 263) adapted from 'Time Profile of Putrescine, Cadaverine, Indole and Skatole in Human Saliva' by M. Cooke, N. Leeves and C. White

'Drowning – Faulkner's most commonly employed means of dispatch for the mules in his work' (p. 269) from 'The Dead Mule Rides Again' by Jerry Leath Mills. Drawing by Bruce Strauch

Appreciating carnivore-livestock conflict (p. 298) adapted from 'Mitigating Carnivore-Livestock Conflict in Europe' by Robin Rigg, Salvomír Findo, Maria Wechelsberger, Martyn L. Gorman, Claudio Sillerno-Zubiri and David W. MacDonald

Figure 1 (p. 300) and Figure 2 (p. 301) adapted from 'Quantification of the Abrasive Wear of a Gold Wedding Ring' by Georg Steinhauser

ABOUT THE AUTHOR

Marc Abrahams is the editor and co-founder of the science humour magazine *Annals of Improbable Research*, a weekly columnist for the *Guardian*, and author of *This Is Improbable: Cheese String Theory, Magnetic Chickens, and Other WTF Research*. He is the founder of the Ig Nobel Prize ceremony, which honours achievements that make people laugh, and then think, every September at Harvard University. Abrahams and the Ig Nobels have been widely covered by the international media, including the BBC, ABC News, the *New York Times*, *Daily Mail*, *The Times*, *USA Today*, *Wired*, *New Scientist*, *Scientific American*, and *Cocktail Party Physics*. In 2012, he symbolically donated Professor Trinkaus and Trinkaus's body of research to BBC Radio 4's *Museum of Curiosity*. Abrahams and his wife, Robin, a columnist for the *Boston Globe*, live in Cambridge, Massachusetts. His ever-burgeoning web site is www.improbable.com.